LEONARDO

LEONARDO DA VINCI
IN HIS OWN WORDS

William Wray

Arcturus Publishing Limited
26/27 Bickels Yard
151–153 Bermondsey Street
London SE1 3HA

Published in association with
foulsham
W. Foulsham & Co. Ltd,
The Publishing House, Bennetts Close, Cippenham,
Slough, Berkshire SL1 5AP, England

ISBN 0-572-03062-2

This edition printed in 2005

British Library Cataloguing-in-Publication Data: a catalogue record for this book is available from the British Library

Design by Elizabeth Healey
Cover Design by Peter Ridley

Printed in Malaysia

Contents

Introduction..6

Leonardo: master artist..................................16

Leonardo's scientific vision.......................100

Leonardo as thinker......................................156

Picture credits..192

INTRODUCTION:

THE GENIUS OF LEONARDO

Leonardo da Vinci is generally recognized as one of the great geniuses of all time – even in his own time he was known as the 'Divine Leonardo'. Vasari, his biographer, a painter in his own right and an astute observer of his contemporaries, made this observation: 'The gifts that Leonardo possessed seemed unlimited, extending to all areas of human knowledge and skill – artist, scientist, architect, musician, engineer, court entertainer, inventor and philosopher.'

Leonardo is generally considered to be the ultimate 'Renaissance man', a man capable of turning his mind to anything. We have his notebooks to prove it, and what we will do is to explore his words in order to come to a better understanding of the man, his achievement and his thinking.

'Learning acquired in youth arrests the evil of old age; and if you understand that old age has wisdom for its food, you will so conduct yourself in youth that your old age will not lack for nourishment.'

RENAISSANCE means 'rebirth', and rebirth for fourteenth-century Florence meant a rediscovery of the learning of ancient Greece and Rome, with Greek texts being translated into Latin, the language of the educated. Leonardo, however, was an artist and therefore trained only in what his craft demanded of him: he certainly was not expected to know Latin, let alone Greek, so he could only study the limited amount of material that was translated into Italian. In a self-deprecating style, Leonardo makes the following observation about his contribution to human understanding and progress:

'Seeing that I cannot find any object of great utility or pleasure, because the men who have come before me have taken for their own all useful and necessary themes, I will do like one who because of his poverty, is the last to arrive at the fair, and not being otherwise able to provide for himself, takes all the things which others have seen and not taken but refused as being of little value.'

What Leonardo offered, that thing 'of little value', was the product of direct observation, and it was this kind of empirical understanding that led to the technological advances that allowed the development of the modern world as we know it.

'To me it seems that all sciences are vain and full of errors that are not born of Experience, mother of all certainty... that is to say, that do not at their origin, middle or end, pass through any of the five senses.'

What Leonardo fully appreciated was that a man does not stand apart from nature, but that he is capable of observation followed by reason and contemplation, and he does have creative power. All of these Leonardo possessed in abundance, as Vasari noted: 'He explained that men of genius sometimes accomplish most when they work least, for they are thinking out inventions and forming in their minds the perfect ideas which they subsequently express and reproduce with their hands.'

Leonardo was not content with a pre-formulated set of assumptions. He was possessed by unlimited curiosity and wanted to lay nature out before his exacting eyes in order to discover what is actually there to be seen. His drawings are wonderful in themselves, but they were linked to his power to express his observations and thoughts in writing and his notebooks contain plenty of writing that is not only analytically precise but also profound in both thought and expression.

Leonardo was determined to discover the unifying principles that ruled all phenomena. He suspected that the principles of proportion that shaped the human body also governed the growth of trees, the flight of birds, and the flow of water. In studying the forces of nature, he intended to show living energy responding to fire, water, earth and air. The mystery of life was the unifying theme of his work,

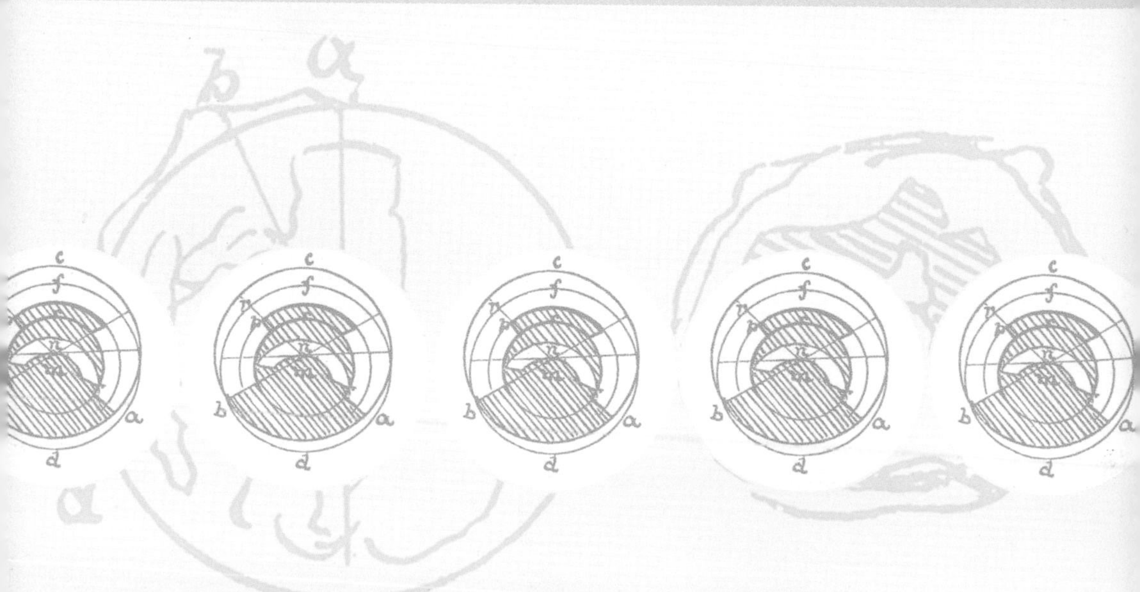

and it is the sense of unity and the sense of depth and mystery that make his paintings so intriguing and powerful.

In the notebook that he carried everywhere on his belt, Leonardo recorded all he observed and the results of his many studies. He intended to write books on a variety of subjects, including painting, human movement, and the flight of birds. None of these were ever completed. What do exist are fragments and first drafts. These fragments, although incomplete, are a goldmine of scientific observation and invention. They indicate Leonardo's thoughts and insights as to the purpose and meaning of life.

By reading Leonardo's words we will look through his eyes, see the world as he saw it and, in the process, discover so much more about the scope of the human mind.

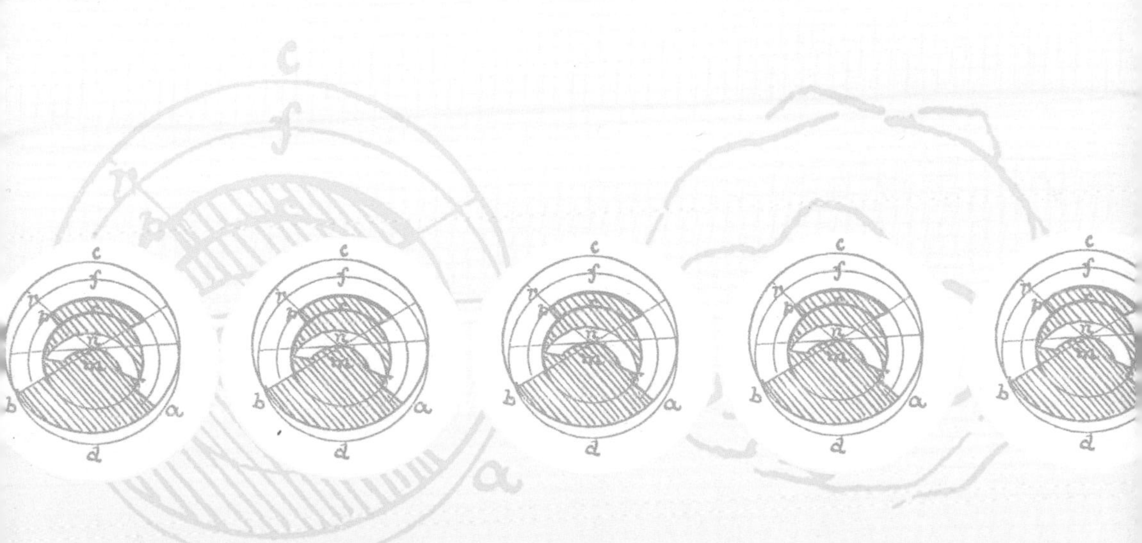

His story

WHEN THE young Leonardo's talent was recognized he was apprenticed to Andrea del Verrochio, head of what was, at the time, Florence's most renowned workshop.

Leonardo was certain that the task of the pupil was to outshine his master. Although Verrochio was a wonderful painter, Vasari tells how Verrocchio abandoned painting to concentrate on sculpture when he saw Leonardo's divine gifts: 'Verrochio was working on a panel picture showing the baptism of Christ by St John, for which Leonardo painted an angel... and despite his youth, he executed it in such a manner that his angel was better than the figures painted by Andrea. This was why Andrea would never touch colours again. He was so ashamed that a boy understood their use better than he did.'

Leonardo spent time at the court of Milan. As a fully trained artist, he was expected to have knowledge of many things, not just how to paint and sculpt. In fact, in his own letter to Duke Lodovico Sforza, the de facto ruler of Milan, he makes

much of his ability as a military engineer; and this was but a small part of what he claimed to be capable of:

> 'I shall contrive catapults, mangonels, 'trabocchi' and other engines of wonderful efficacy... In short, to meet the variety of circumstances, I shall contrive various and endless means of attack and defence.'

Whether it was because he couldn't find a patron or because his questing mind would not allow him to rest, Leonardo spent some time in the service of the infamous Cesare Borgia. Leonardo had by now taken up the study of topography and geology and this served him well, as Borgia made him his chief engineer and sketches for military installations from this time are to be found in his notebooks.

Leonardo was commissioned by the Florence's city council to paint a large fresco on one of the walls of the new Sala di Gran Consiglio in the Palazzo Vecchio. Leonardo chose to portray the Battle of Anghiari, in which Florence had defeated Milan in 1440. His rendition showed a cavalry battle, and, breaking completely from convention, the picture is full of violent movement as the figures of men and horses clash and tumble over one another.

'Make the conquered beaten and pale, with brows raised up and knit, and the skin above their brows furrowed with pain... Show someone using one hand for a shield for his terrified eyes with palm turned towards the enemy.'

It was whilst he was working on *The Battle of Anghiari* that Leonardo painted the *Mona Lisa*, perhaps the most renowned painting in the world. There is no doubt that the harmony that the painting exudes is a key feature of Leonardo's work.

Leonardo's scientific approach to the aim of perfect depiction of all things led him inevitably to the need for understanding the physical insides of things. To this end, the famous anatomist Marc Antonio della Torre helped him in his researches and Leonardo produced this note – and a wonderful drawing, presumably from an autopsy – of a child in the womb:

'It lies continually in water, and if it were to breathe it would be drowned, and breathing is not necessary to it since it receives life and is nourished from the life and food of its mother.'

Leonardo's old age was assured when his patron, Giuliano Medici, died. Leonardo, feeling old and neglected noted that '*the Medicis made me...and ruined me*'. He needn't have complained, however, for Francis I of France persuaded Leonardo to go to France, settling him in the small castle at Cloux, near the royal palace at Amboise on the Loire.

Francis had a genuine love of art and learning, and naturally enough he was an admirer of Leonardo: 'No other man had been born who knew as much about sculpture, painting and architecture, but still more he is a very great philosopher.'

Leonardo was nearing the end of his life, and on the 24th June 1518 he died. In his biography Vasari included this assessment of the man's genius: '... occasionally, in a way that transcends nature, a single person is marvellously endowed by heaven with beauty, grace and talent in such abundance that he leaves other men far behind...Everyone acknowledged that this was true of Leonardo da Vinci.'

We may judge the truth of Vasari's assessment from what Leonardo left to us: his paintings, his drawings and a record of his thoughts and words in his notebooks.

LEONARDO: MASTER ARTIST

'O thou that sleepest, what is sleep? Sleep is an image of death. Oh why not let your work be such that after death you become an image of immortality?'

One of the functions of art is to allow us to escape from, and challenge, the narrowness of our own vision and discover something of the breadth and depth, beauty and wholeness of life. Regardless of what our own opinion might be about the nature and purpose of art, there is no doubt about the power of Leonardo's paintings, drawings and sketches.

'Every part is disposed to unite with the whole, that it may thereby escape from its own incompleteness.'

Leonardo thought hard and long about his role as artist. He wanted far more for the artist than had ever been attained before. In our contemporary world and thinking, the artist has an exulted position as seer and visionary. In Leonardo's time the artist had a far more humble role: he was a craftsman producing work to order in a variety of forms. He would toil not alone but in workshops with other skilled craftsmen and their apprentices, producing not just paintings but all kinds of artefacts.

The 'Seven Liberal Arts' governed the traditional order of scholastic study throughout the Medieval period. Amongst these seven, poetry and music were included, with painting categorised in the lowly position of 'Mechanical Arts', which also included manual labour. The painters of the Renaissance changed all that and Leonardo, in particular, made the case for the artist.

'Whatever exists in the universe, in essence, in appearance, in the imagination, the painter has first in his mind and then in his hand; and these are of such excellence that they present a proportioned and harmonious view of the whole, that can be seen simultaneously, at one glance...'

The fact that Leonardo lived at the time of the Renaissance was a distinct advantage. It meant that with the rediscovery of the Classical world there was a rediscovery of an altogether different role for the artist: to transform man's understanding of himself. The Classical philosopher, Plato, put it well: 'Let our artists rather be those who are gifted to discern the true nature of the beautiful and the graceful; then will our youth dwell in the land of health, amid fair sights and sounds, and receive the good in everything: and beauty; the effluence of fair works, shall flow into the eye and ear, like a health giving breeze from a purer region, and insensibly draw the soul from earliest years into the likeness and sympathy with the beauty of reason.'

Leonardo made considerable claims for the power of art. He described it as a science that endowed passing beauty with a permanence greater than nature. He claimed that:

'Such a science is in the same relation to divine nature as its works are to the works of nature.'

Inherent in this thinking is that nature is not so much physical as divine. Leonardo the scientist believed that it was mankind's purpose to discover the laws of nature that ruled nature's manifestation and the true artist should therefore go beyond the mere copying of nature and come to a direct understanding of it.

'[The artist] should not seek distraction in company but live a life of complete harmony with the natural world and in the process to penetrate the outer forms of nature and discover something of its inner core.'

This was achieved, as is evident in his work, by an empirical investigation of nature in order to discover how nature 'functioned'. If mankind was able to discover, through reflection and study, something of the way the Creator created, Leonardo believed that he was in a far better position to be creative himself.

To achieve this he developed a full programme of study for his apprentices. Although this was never properly structured in his notebooks, there is no doubt that it was entirely systematic. Judging from his writings, the painters that Leonardo instructed had to know how the eye functioned and the mathematical laws that allowed the mind to judge distance. They also had to know how to reproduce a sense of three-dimensional space and so not only did they have to understand the laws of optics, they also had to understand how light modelled all that they saw. They had to understand colour, the effect that one colour has on another and how atmosphere altered colour.

Above all they had to understand proportion and harmony; how the whole complexity of life related as a unified and harmonious whole. Only out of this detailed and complete understanding could the artist, in Leonardo's estimation, be considered a true master.

> 'In truth great love springs from the full knowledge of the thing that one loves; and if you do not know it you can love it but little or not at all.'

This is the course of study that Leonardo put himself through, and there can be no doubt as to the kind of master that he himself was, and remains today.

The painter and painting

LEONARDO was a master of all the techniques that he encouraged his students to follow.

'The painter is lord of all types of people and things... If the painter wishes to see monstrosities that are frightful and bufoonish or ridiculous, or pitiable he can be lord and god thereof... if he wants from high mountain to unfold a great plain extending down to the sea's horizon, he is lord to do so.'

'... whatever exists in the universe, in essence, in appearance, in the imagination, the painter has first in his mind and then in his hand.'

'He who despises painting loves neither philosophy or nature. If you despise painting, which is the sole imitator of all the visible works of nature, you will be certainly despising a subtle invention which brings philosophy and subtle speculation to bear upon the nature of all forms – sea and land, plants and animals, grasses and flowers...'

The seeing eye

LEONARDO'S observations as recorded in his notebooks are anything but ordered, and yet it is evident that if he had ever managed to structure his notes, he would have begun any book on the training of the artist with that which is utterly fundamental to the painter; the use of their eyes.

'Here forms, here colours, here the character of every part of the universe are concentrated to a point; and that point is so marvellous a thing... Oh! Marvellous, O stupendous Necessity – by thy laws thou dost compel every effect to be the direct result of its cause, by the shortest path. These indeed are miracles.'

'Beh[illegible]ere O reader! A thing
conc[illegible] can at
any [illegible]
expe[illegible]ges
unk[illegible] The eye,
who[illegible]know by
expe[illegible]n time,
bee[illegible]ber of
auth[illegible] by
exp[illegible]her.'

The wonder of sight

WITH HIS drawings of the working of the optic nerve, Leonardo wrote about the five senses being connected to the 'organ of perception', the *sensus communis*, where the senses met in one particular spot directly behind the eyes, and which was the seat of the soul.

'Who would believe that so small a space could contain the images of all the universe?'

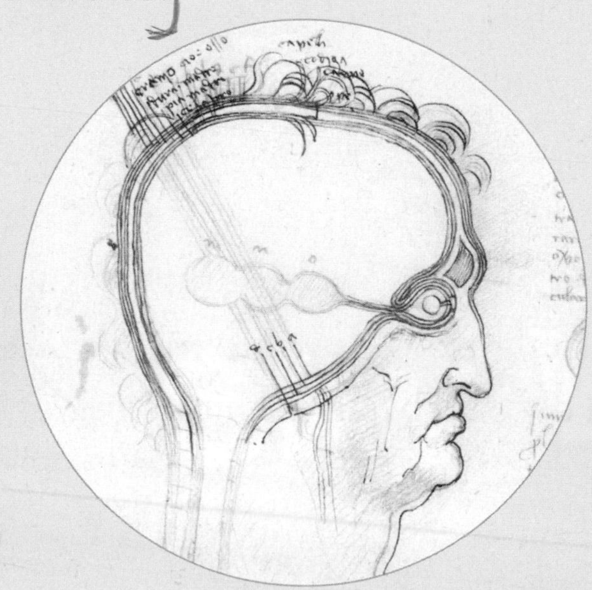

Opposite: *St Bartholomew*

23

'The sense which is nearest to the organ of perception functions most quickly; and this is the eye, the chief and leader of all the others; of this only will I treat and leave the others in order not to be too long.'

'The eye is the window of the human body through which it feels its way and enjoys the beauty of the world.'

'Describe in thy anatomy what proportion there is between the diameters of all lenses in the eye and the distance from these to the crystalline lens.'

It wasn't just their physical immediacy to the seat of the soul that gave the eyes their importance. In Leonardo's estimation, sight was the superior sense in every respect.

'Now do you not see that the eye embraces the beauty of the whole world? It counsels and corrects all the arts of mankind... it is the prince of mathematics, and the sciences founded on it are absolutely certain. It has measured the distances and sizes of the stars it has discovered the elements and their location... it has given birth to architecture and to perspective and to the divine art of painting.'

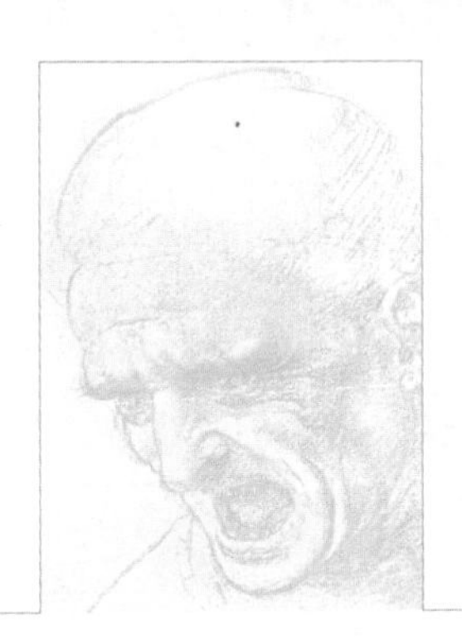

Divine light

BEING A MASTER of the representation of the play of light, Leonardo had closely observed that which stimulated the eye. He understood that light was a physical element, but he also interpreted it figuratively as the light of the mind and the light of the spirit.

'Light is the chaser away of darkness. Look at light and consider its beauty. Blink your eye and look at it again: what you see was not there at first, and what was there is no more.'

He understood from experience how the artist stood in contemplation at the point where that inner and outer light met.

'Even as the Lord who is the light of all things shall vouchsafe to enlighten me I will treat of light.'

'Among the various studies of natural processes, that of light gives most pleasure to those who contemplate it.'

'All surfaces of solid bodies turned towards the sun or towards the atmosphere illuminated by the sun become clothed and dyed by the light of the sun or the atmosphere.'

'A single and distinct luminous body causes stronger relief in the objects than a diffused light; as may be seen by comparing one side of a landscape illuminated by the sun, and one overshadowed by clouds, and illuminated only by the diffused light of the atmosphere.'

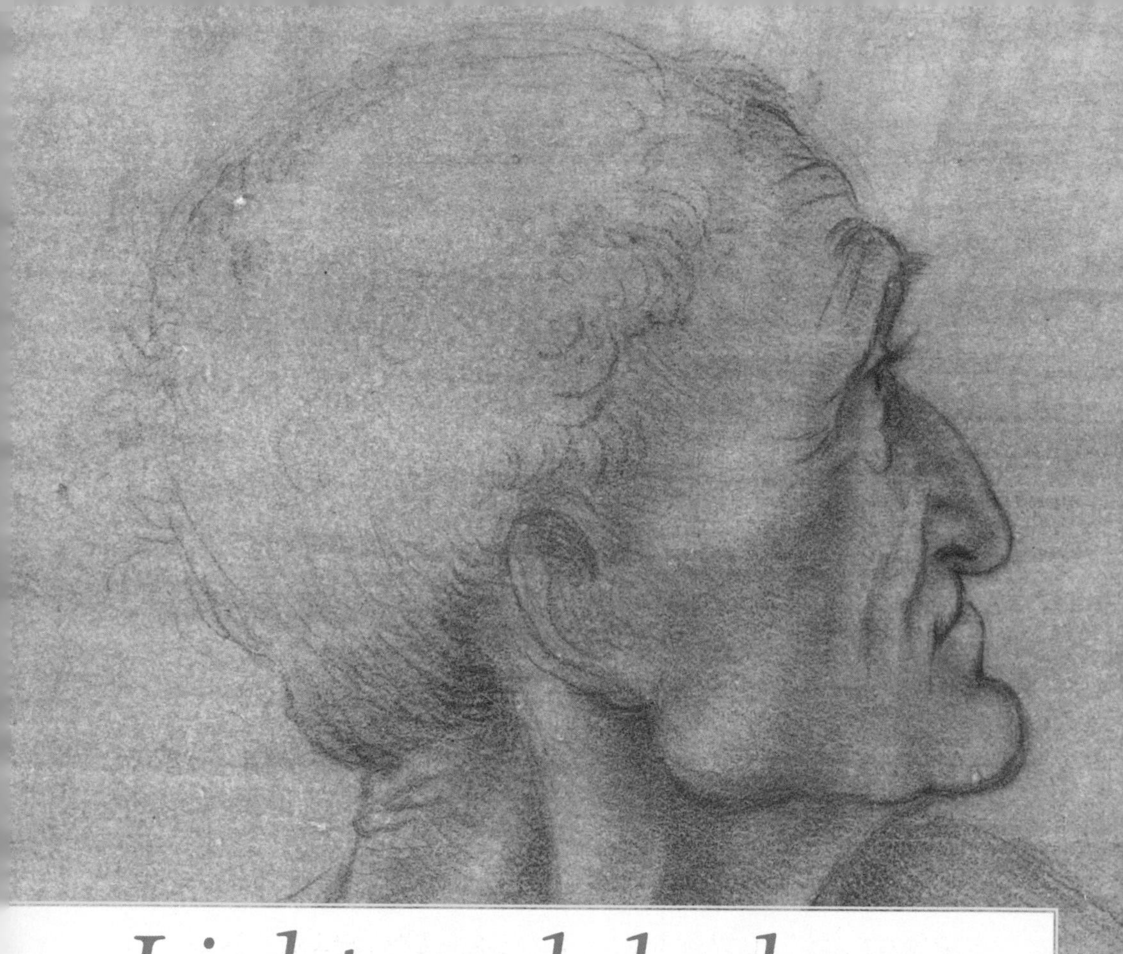

Light and darkness

In Leonardo's paintings are to be seen the infinite gradations of light and dark.

'Every solid body is surrounded and clothed with light and darkness. You will get only a poor perception of the details of a body when the part that you see is all in shadow, or all illuminated.'

He understood this play as no other painter had done before him, displaying his expertise in the techniques of *chiaroscuro* and *sfumato*.

‘The more brilliant the light of a luminous body, the deeper the shadows cast by the illuminated object.’

‘In an object in light and shade, the side which faces the light transmits the images of its details more distinctly and immediately to the eye than the side which in shadow.’

In his record of nature, Leonardo was the master of the brilliance of illumination contrasting with tonal gradation into mystery and darkness.

'Very great charm of shadow and light is to be found in the faces of those who sit in the doors of dark houses. The eye of the spectator sees that part of the face which is in shadow lost in the darkness of the house, and that part of the face which is lit draws its brilliancy from the splendour of the sky. From this intensification of light and shade the face gains greatly in relief and beauty by showing the subtlest shadows in the light part and the subtlest lights in the dark part.'

Opposite: *St Anne with the Virgin and Child, and John*

Drapery

ONE OF the most obvious representations of the play of light and dark is shown in the drawing of drapery. There are, in his portrayal of the softly modelled fall of cloth, vivid examples of all that Leonardo wrote about the contrast between light on the high point of a fold and the deep darkness in the depth of the crease.

'You must not give drapery a great confusion of many folds, but rather introduce only those held by the hands or arms; the rest you may let fall simply where its nature makes it flow.
And the folds should correspond to the quality of the draperies, whether transparent or opaque...'

In his notebooks Leonardo wrote about the appropriateness of different kinds of cloth to different characters:

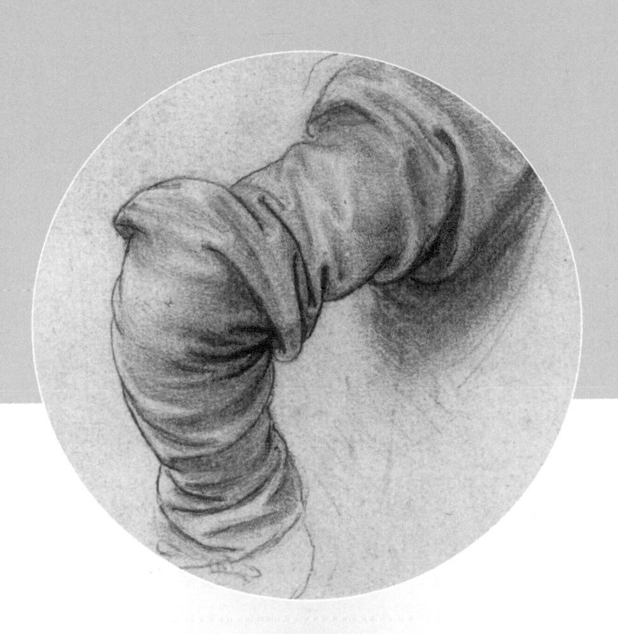

'The limbs of a nymph or an angel should be shown in almost their original state, for these are represented clad in light draperies which are driven and pressed against the limbs of the figures by the blowing of the wind.'

'The draperies – thin, thick, new, old, with folds broken and pleated, soft light, shadows obscure and less obscure, with or without reflections, definite or indistinct according to distances and colours...'

Light and colour

LEONARDO TOOK close note of the way that light created colour.

'Since we see that the quality of colour is seen only by means of light, it is to be supposed that where there is most light the true character of a colour in light will best be seen; and where there is most shadow the colour will be affected by the tone of that. Hence, O Painter! Remember to show the true quality of colours in bright lights.'

Spilling onto everything are the colours reflected by nearby objects and it was this infinitely subtle interaction that Leonardo conveyed in his paintings.

'You may have seen how the light that penetrates through stained glass windows in churches assumes the colour of the glass of these windows.'

Opposite: *St Anne with the Virgin and Child*

'Watch the sun as it's setting when it appears red through the vapour, and dyes all the clouds that reflect its light.'

'The colours in the middle of the rainbow mingle together. The bow in itself is not in the rain nor in the eye that sees it; though it is generated by the rain, the sun and the eye.'

'If you are representing a white body let it be surrounded by ample space, because as white has no colour of its own it is tinged and altered in some degree by the colour of the objects surrounding it. If you see a woman dressed in white in the midst of a landscape that side which is towards the sun is bright in colour, so much so that in some proportions it will dazzle the eyes like the sun itself; and the side which is towards the atmosphere, luminous through being interwoven with the sun's rays and penetrated by them... that side of the woman's figure will appear steeped in blue.'

The power of the eyes

IN THE Renaissance, with its stress on man's own innate power, the notion of the eye as a projector was discussed. One of the ideas circulating amongst the Neo-platonic thinkers of Florence was that of the *idolum*, the power that all things possess to give off both their physical shape and their inner energy. Leonardo toyed with the idea that seeing was not just a passive activity, but that one had to send out visual rays from the eye, in order to capture the images which pervaded the surrounding air.

'The eye transmits its own image through the air to all the objects which face it, and also receives them on its own surface, whence the "sensus communis" takes them and considers them.'

'I hold that the invisible powers of imagery in the eyes may project themselves to the object as to the images of the objects to the eyes.'

'I say that the power of vision extends through the visual rays to the surface of non-transparent bodies, while the power possessed by these bodies extends to the power of vision.'

'Maidens are said to have power within their eyes to attract to themselves the love of men.'

Opposite: Angel from *The Madonna of the Rocks*

Perspective

For Leonardo, the laws of perspective – the organized method of representing a three-dimensional world on a two-dimensional surface – formed the foundation upon which the whole of the painter's creation arose. His understanding of perspective was three-fold:

'The first deals with the reasons of the (apparent) diminution of objects as they recede from the eye, and is known as Perspective of Diminution; the second contains the way which colours vary as they recede from the eye; the third and last explains how objects should appear less distinct in proportion as they are more remote. And the names are as follows: Linear perspective, the perspective of colour, the perspective of disappearance.'

Leonardo was utterly familiar with the work of Filippo Brunelleschi and Leon Battista Alberti, both great men of the Renaissance, and his knowledge of their findings, in conjunction with his anatomical study of the eye and his work on optics, made him a complete master of perspective.

'Among objects of equal size, that which is most remote from the eye will look the smallest.'

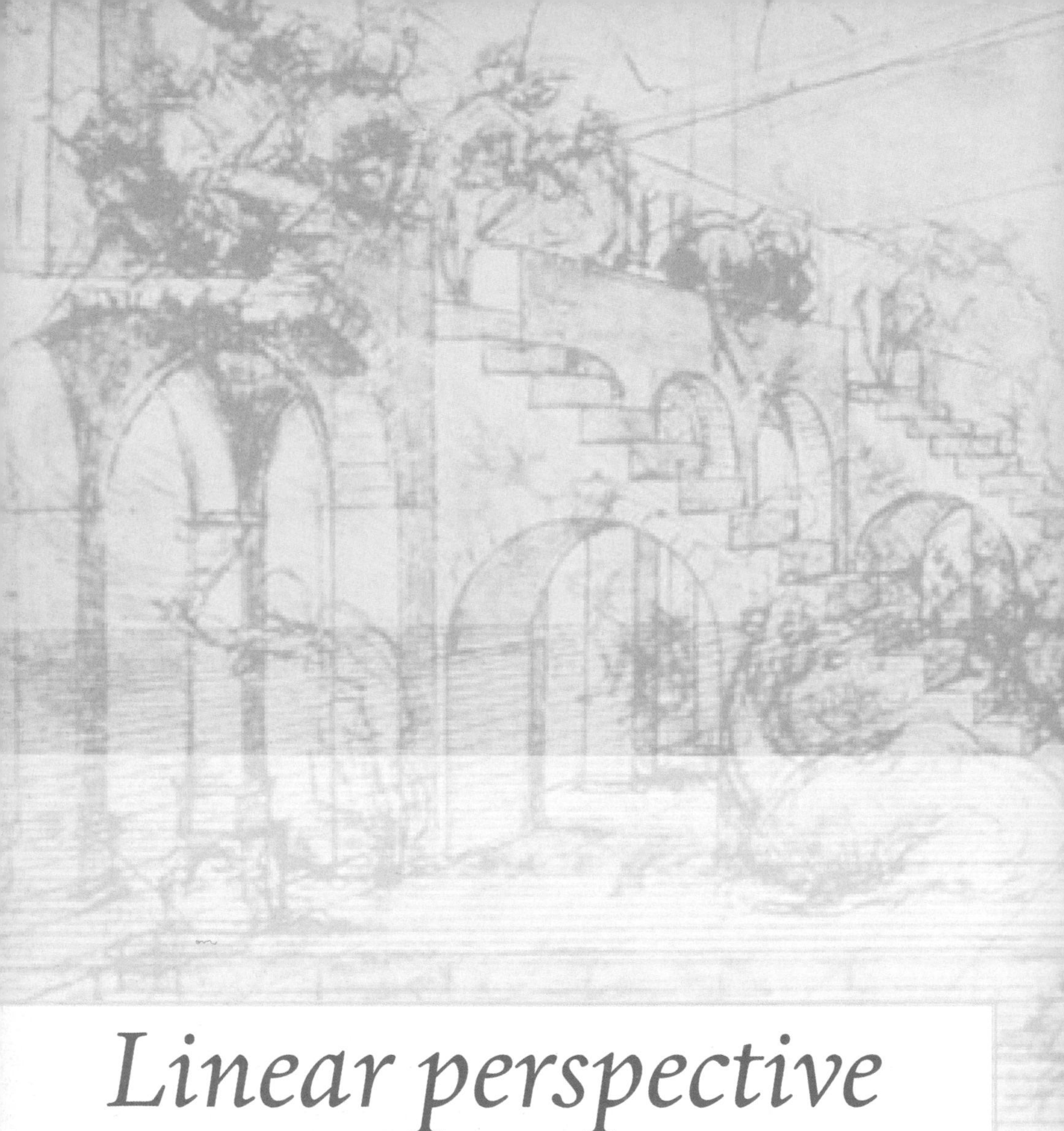

Linear perspective

LINEAR PERSPECTIVE is based on the impression that parallel lines appear to converge as they recede to what is called the 'vanishing point'. In Leonardo's notebooks there is a translation of John Pecham's *Perpectiva Communis*, a medieval handbook on optics whose influence remained undiminished throughout the Renaissance.

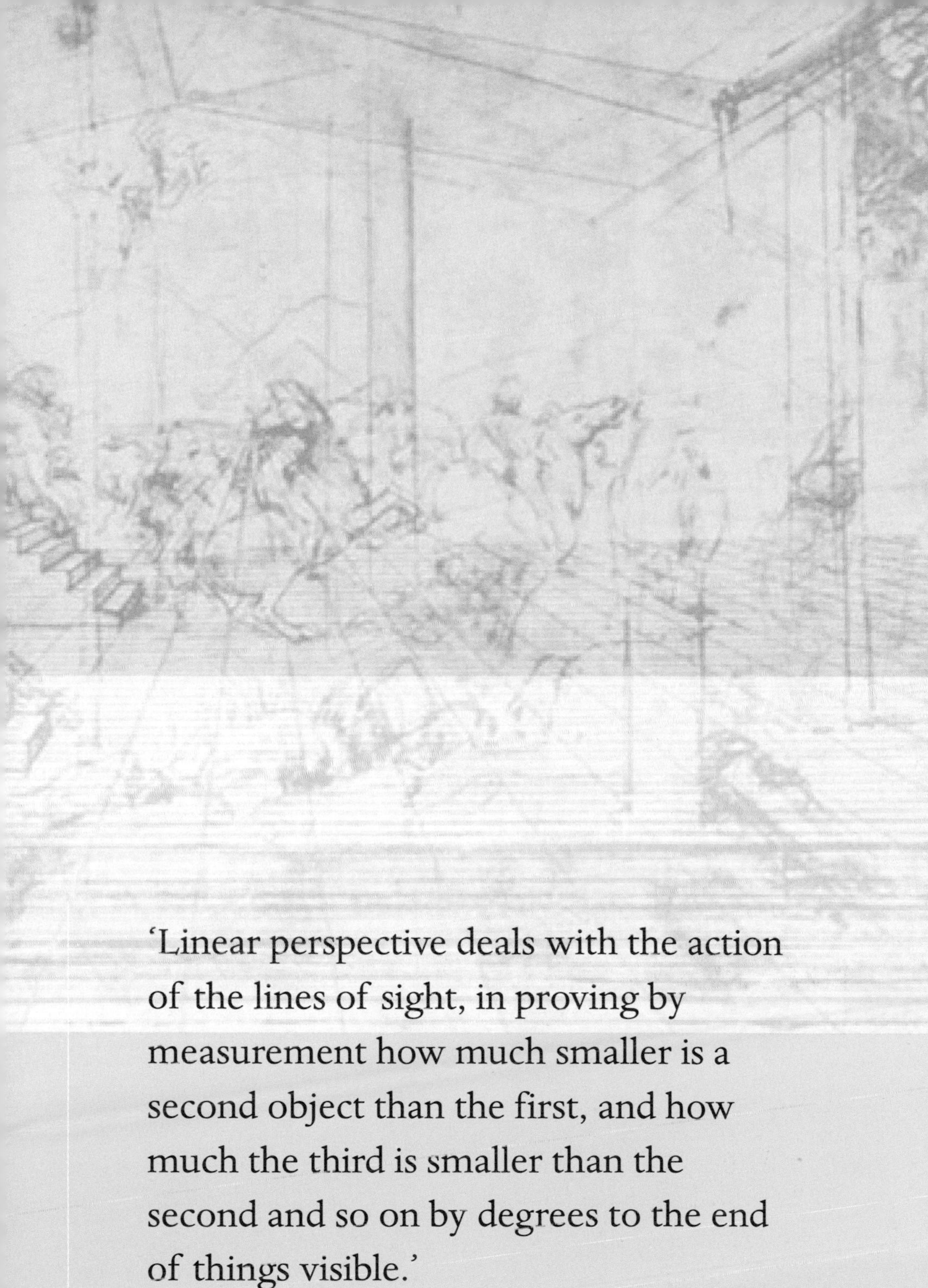

'Linear perspective deals with the action of the lines of sight, in proving by measurement how much smaller is a second object than the first, and how much the third is smaller than the second and so on by degrees to the end of things visible.'

Above: Sketch for *The Adoration of the Magi*

Perspective is the key to the presentation of what appears to be three dimensions on a two-dimensional surface. It is achieved by an examination of the optical laws which govern the way the eyes read the world, recognize distance and appreciate depth. In medieval pictures, the observer looked at a flat surface full of signs and symbols derived from the natural world. Presented in Renaissance paintings, it was the natural world itself that was appreciated; a world in which it was seemingly possible to move beyond the picture frame and actually 'enter' the scene the picture portrayed.

'Perspective is a rational demonstration which by experience confirms the images of all things sent to the eye by pyramidal lines.'

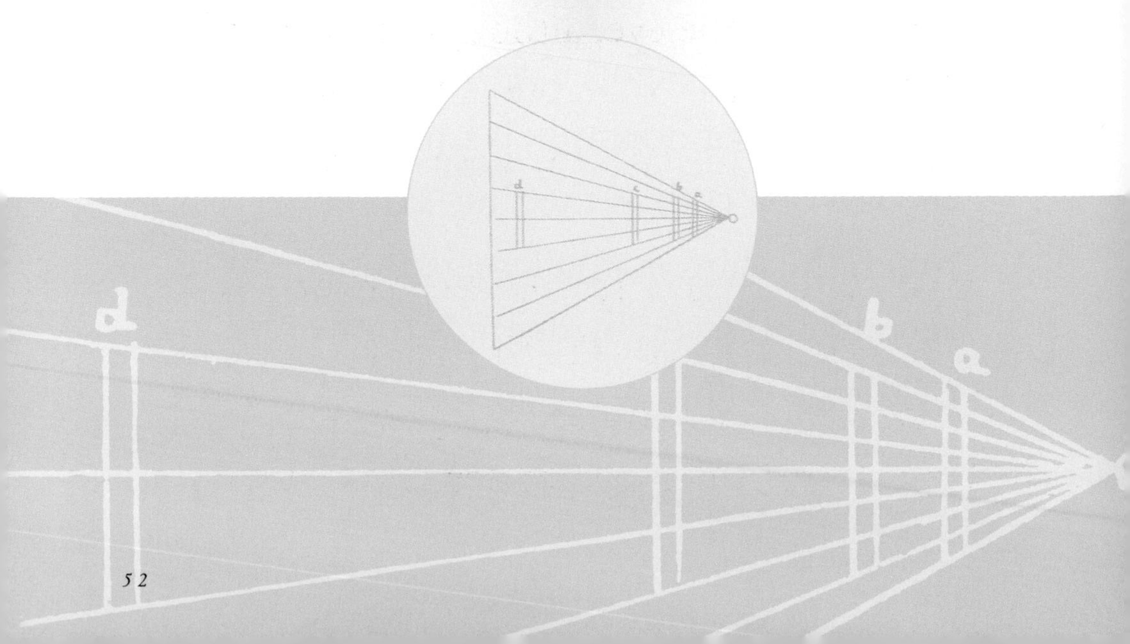

'The air is full of straight and radiating lines intersected and interwoven with one another. They represent too whatever object the true form of their cause.'

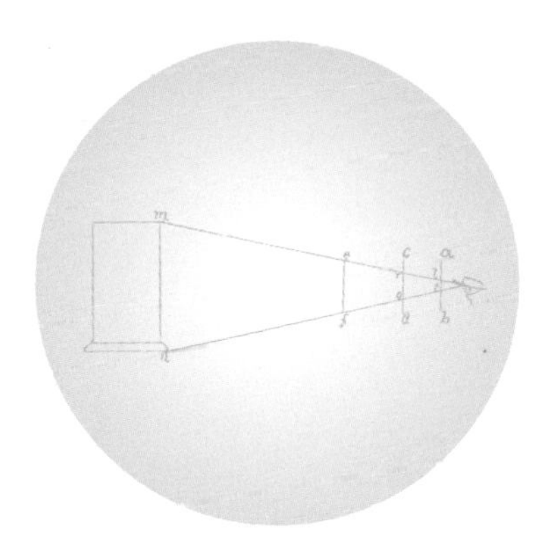

'I find by experience that if a second object is as far beyond the first as the first is from the eye, although they are of the same size the second will seem half the size of the first, and if the third object is of the same size as the second, and the third is as far beyond the second as the second from the first, it will appear of half the size of the second, and so on by degrees, at equal distances.'

Aerial perspective

LEONARDO SAW that the density of the atmosphere made a considerable difference to how distant things appeared. One of the things that Leonardo is famed for is his mastery of *sfumato*, which meant painting with minute gradations of tone to produce a misty or vaporous effect. He used this technique as a way of conveying aerial perspective.

'If in painting you wish to represent one [*object*] more remote than another you must make the atmosphere somewhat dense.'

'The painter can suggest to you various distances by a change of colour produced by the atmosphere intervening between the object and the eye.'

He could depict mists through which the shapes of things could only be discerned with difficulty; rain with cloud-capped mountains and valleys showing through; clouds of dust whirling about combatants; streams of varying transparency, and fishes at play between the surface of the water and its bottom.

'Of several bodies all equally larger and distant, that most brightly illuminated will appear to the eye nearest and largest.'

'Of several bodies of equal size and tone, that which is farthest will look lightest and smallest.'

'Of shadows of equal depth, those nearest the eye will look least deep.'

Landscape

LEONARDO recommended that artists walked alone in the countryside so as to appreciate all the more keenly the beauty of nature. His own aesthetic appreciation, coupled with his work as a civil engineer, required an intimate knowledge of the land and enabled him to portray landscape with exquisite understanding.

'Landscape is of a more beautiful azure when in fine weather the sun is at noon, than at any other time of the day, because the air is free from moisture.'

'The colours of shadows in mountains at a great distance take a most lovely blue, much purer than their illuminated portions.'

'When the sun is covered by clouds, objects are less conspicuous, because there is little difference between the light and shade of the trees and the buildings being illuminated by the brightness of the atmosphere which surrounds the objects in such a way that the shadows are few, and these few fade away so that their outline is lost in haze.'

Trees in a landscape

As part of Leonardo's study of the landscape, he made extensive notes on painting trees.

'The accidental colour of trees are of four, namely shadow, light, lustre and transparency.'

'If you are on the side whence the wind is blowing you will see the trees looking much lighter than you would see them on the other sides; and this is due to the fact that the wind turns up the reverse side of the leaves which in all trees is much whiter than the upper side.'

Leonardo had a great admiration for trees, greatly appreciating their beauty and the way in which they interacted with the light.

'The trees in landscape which are between you and the sun are far more beautiful than those which have you between the sun and themselves, and this is so because those which are in the same direction as the sun show their leaves transparent towards their extremities and the parts that are not transparent, that is at the tips, are shining; and the shadows are dark...'

'The part of the tree, which has shadow for background, is all of one tone, and whenever the trees or branches are thickest thcy will be darkest, because there are no little intervals of air.'

Botany for painters

THE NOTEBOOKS of Leonardo contain exquisite botanical drawings, and scattered about his paintings are a wide variety of easily recognizable plants.

'A leaf always turns its upper side towards the sky so that it better receive, on all its surface, the dew, which drops gently from the atmosphere.'

'Nature is so delightful and abundant in its variations that there would not be one that resembles another, and not only plants as a whole, but among their branches, leaves and fruit, will not be found one which is precisely like another.'

Leonardo's botanical interest lasted throughout his life, with studies on the growth and forms of plants regularly appearing in his notebooks.

'The sun gives spirit and life to the plants and the earth nourishes them with moisture.'

'The ramifications of any tree, such as the elm, are wide and slender after the manner of the hand with spread fingers foreshortened.'

Wind, sky and air

WITH HIS close study of the play of the elements and his understanding of the light of the sun as it passed through the earth's atmosphere, Leonardo inevitably devoted a number of observations to wind and weather, of clouds up against the sun and of wind shaking leaves or bending boughs.

'Describe landscapes with the wind, and the water and the setting and rising of the sun.'

'The shadows in the clouds are lighter in proportion as they are nearer the horizon.'

'When the clouds come between the sun and the eye all the edges of their round masses are light and they are dark towards the centre...'

'The black clouds which are often seen higher up than those which are illuminated by the sun are shaded by other clouds, lying between them and the sun.'

'The rounded forms of the clouds that face the sun, show their edges dark because they lie against a light background; and to see this is true, you may look at the top of any cloud that is wholly light because it lies against the blue of the atmosphere, which is darker than the cloud.'

Proportion

Among the notes that he prepared for a proposed book *The Artist's Course of Study*, Leonardo discussed the importance of proportion in the body. Vitruvius, the renowned Roman architect, considered that the proportions of the human body related to fundamental geometric principles. It was an opinion with which Leonardo concurred, and he drew his own version of the 'Vitruvian Man':

'... 4 fingers make 1 palm; 4 palms make 1 foot; 6 palms make one cubit; 4 cubits make a man's height; and 4 cubits make one pace; and 24 palms make a man.'

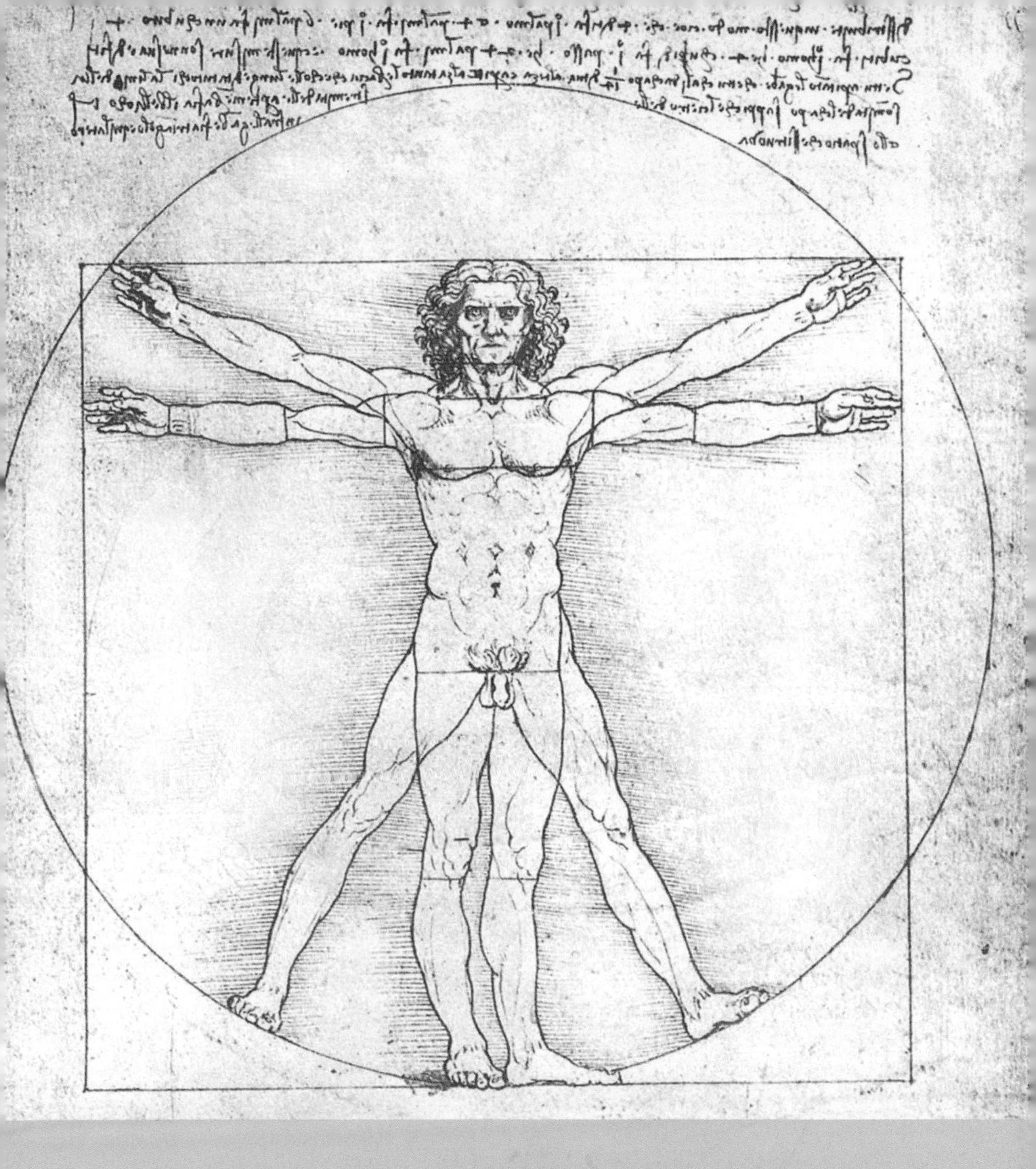

'Man is the measure of all things… Every part of the whole must be in proportion to the whole… I would have the same thing understood as applying to all animals and plants.'

Leonardo spoke in detail about the importance of creating harmonious concord, and studied the proportions of the various limbs.

'If you open your legs so as to decrease your height by one fourteenth and spread and raise your arms so that your middle fingers are on a level with the top of your head, you must know that the navel will be the centre of the circle of which the outspread limbs touch the circumference; and the space between the legs will form an equilateral triangle.'

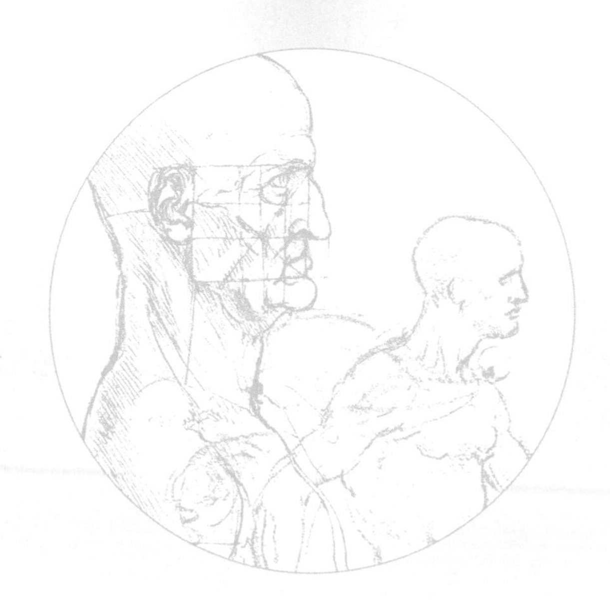

'The span of a man's outstretched arms is equal to his height.'

'Every man at three years is half the full height he will grow to at last.'

The human body

LEONARDO was a dedicated anatomist, exploring the functioning of the body by meticulous dissection. He was aware of the body as an exquisite piece of engineering, and very conscious of how the movement of the limbs were achieved through the contraction and expansion of muscles and tendons.

'The painter who has a knowledge of the sinews, muscles and tendons will know very well in the movement of a limb, how many and which of the sinews are the cause of it, and which muscle by the swelling is the cause of the contraction of that sinew.'

In his advice to students, Leonardo pointed out areas to look for in the portrayal of the human body at all ages, in a whole range of conditions.

'Thus a youth has limbs that are not very muscular nor strongly veined, and the surface is delicate and round and tender in colour. In a man the limbs are sinewy and muscular; while in old men the surface is wrinkled, rugged and knotty, and the veins very prominent.'

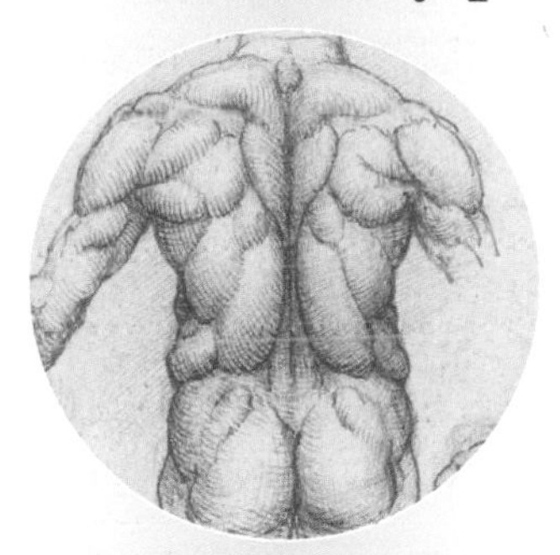

'Children are slender at the joints and fat between the joints; as may be seen in the joints of the fingers, arms and shoulders, which are slender and dimpled, while in man on the contrary all the joints of fingers, arms and legs are thick.'

The body in motion

IN HIS USUAL analytical fashion, Leonardo identified different kinds of movements, and what to consider when capturing the essence of those movements.

'The faster a man runs, the more he leans forward towards the point he runs to.'

'If you wish to produce a figure that shall look light and graceful in itself you must make the limbs elegant and extended without too much display of muscles.'

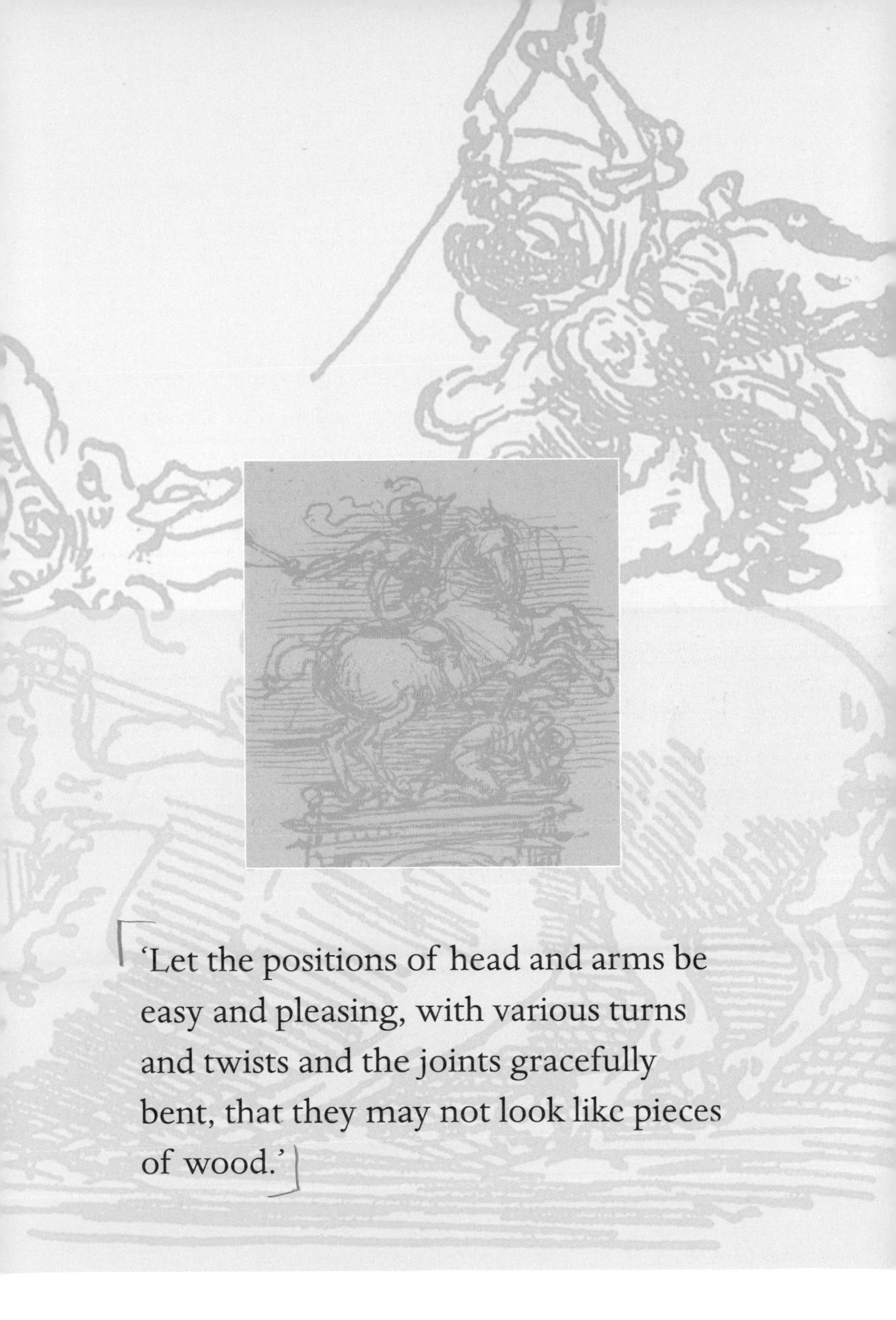

'Let the positions of head and arms be easy and pleasing, with various turns and twists and the joints gracefully bent, that they may not look like pieces of wood.'

Faces

WHEN LEONARDO talked of the inner nature of 'man' he was not only speaking of anatomy; he was also referring to the subtle movement of mind and spirit. His portraits were perhaps the greatest masterpieces of this inner life. Of the qualities portrayed in his pictures he said very little, realizing perhaps that anything he might have said could in no way compare with what he managed to convey in paint.

We are left with what others had to say about Leonardo as a portraitist. Vasari's comments on the *Mona Lisa* are apt: 'The eyes had their natural lustre and moistness, and around them were the lashes and all those rosy and pearly tints that demand the greatest delicacy of execution. The eyebrows were completely natural, growing thickly in one place and lightly in another and following the pores of the skin. The nose was finely painted, with rosy and delicate nostrils as in life. The mouth joined to the flesh-tints of the face by the red of the lips, appeared to be living flesh rather paint. On looking closely at the pit of her throat one could swear that the pulses were beating. Altogether this picture was painted in a manner to make the most confident artist – no matter – despair and lose heart.'

Opposite: *The* Mona Lisa

Leonardo, however, did record this heartfelt comment:

'The proportions of the beautiful forms that compose the divine beauties of this face here before me, which being all joined together and reacting simultaneously give me much pleasure with their divine proportions that I think there is no other work of man on earth that can give me greater pleasure.'

Advice to painters

LEONARDO wrote at length about the techniques that a 'master' artist should have available, and was certain of one thing:

'He is a poor painter who does not excel his master.'

Leonardo was to have written a book for students of art, *The Artist's Course of Study,* but all we have of it is his notes. In these, he reflected on the nature of the artist's life: how to avoid distraction; how to make a full connection with the world about him; and how to combine observation, reason and the imagination to produce something that was not merely a copy of the work of the masters of the past.

'Which is best, to draw from nature or from the antique?'

In Leonardo's estimation the artist had to return constantly to the true master, 'nature itself.' Painting enabled one to come to know and love 'the maker of all marvellous things'.

Opposite: Detail from *The Adoration of the Magi*

'First draw from drawings by good masters... then from plastic work, with the guidance of the drawing done from it; and then from good natural models.'

'... always go slowly to work in your drawing, and discriminate.'

The solitary life

LEONARDO WAS never short of friends, but such was the nature of the man, so far in advance was he of current thinking, that few if any properly understood him. He stood apart, and evidently considered that the solitary life was a positive asset.

'The painter ought to be solitary and consider what he sees, discussing it with himself in order to select the most excellent parts of whatever he sees.'

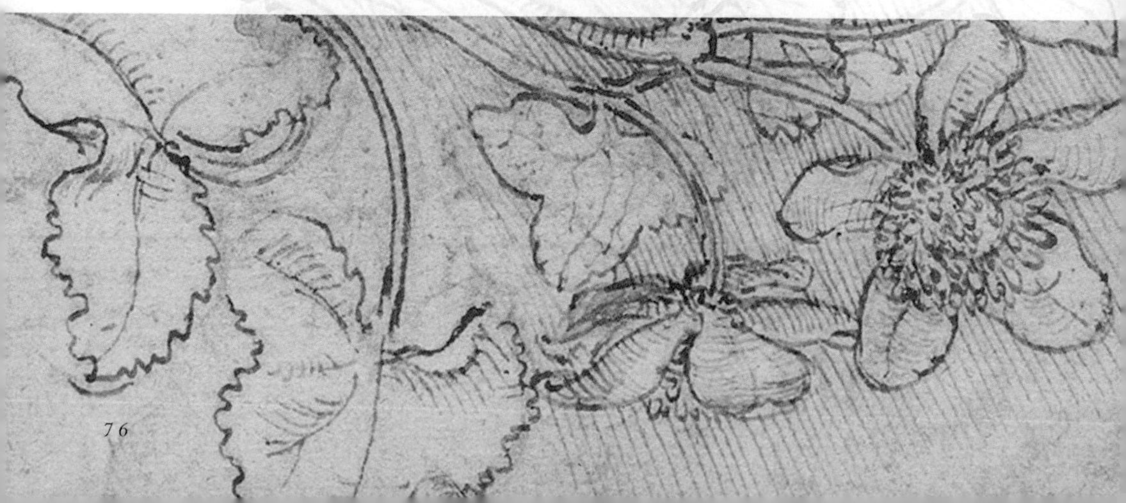

'While you are alone you are entirely on your own; and if you have but one companion you are but half your own, or even less in proportion to the indiscretion of his conduct.'

Only by being solitary could one achieve an unfettered connection with the world in all its beauty.

'What induces you, oh man, to depart from your home in town, to leave parents and friends, and go to the countryside over mountains and valleys, if it is not for the beauty of the world of nature?'

Holding a mirror up to nature

LEONARDO BELIEVED the task of the artist was firstly to hold the mirror up to nature, to note in all its detail the variety of nature.

'He should act as a mirror which transmutes itself into as many colours as are those of the objects that are placed before it... Above all he should keep his mind as clear as the surface of a mirror.'

For the artist to observe accurately, Leonardo was certain that not only close attention was required but also a still and clear mind.

'As you walk through the fields turn your attention to the various objects and look now at this thing and now at that collecting a store of divers facts.'

'Do you not see what variety of mountainous regions and plains, of springs, rivers, cities with public and private buildings, instruments fitted for man's use; of different costumes, ornaments and arts? All these should be rendered with equal facility and perfection by whomever you wish to call a good painter.'

Reason and observation

THE MIRROR of the mind is not merely a passive receptor. As we have already explored, Leonardo was a scientist, intent on discovering the underlying laws that governed creation.

'The painter who draws by practice and judgement of the eye without the use of reason is like a mirror which copies everything placed in front of it without knowledge of the same.'

Leonardo was also firmly convinced that the whole process of observation should also be understood.

'Good judgement proceeds from clear understanding, and a clear understanding comes from reason derived from sound rules, and sound rules are the daughters of sound experience – the common mother of all the sciences and arts.'

'Practice should always be based on sound theory, of which perspective is the guide and gateway, and without it nothing can be done well in any kind of painting.'

Intimate knowledge

LEONARDO WAS intent on using both mind and heart in his understanding and expression of the glory of nature. He was not content simply to wonder. However, his objective observation was not merely a matter of cold analysis; his intimate knowledge arose from profound love of all that nature offered him.

'And you, O Man, who will discern in this work of mine the wonderful works of Nature, if you think it would be a criminal thing to destroy it, reflect how much more criminal it is to take the life of a man...'

'Why does the eye see a thing more clearly in dreams than with the imagination being awake?'

'... we might say that the earth has a spirit of growth; that its flesh is the soil, its bones the arrangement and connection of the rocks of which the mountains are composed, its cartilage the tufa, and its blood the springs of water.'

Opposite: The Head of A Woman

People in the street

As a close observer of the world in all its variety, Leonardo was in touch with all he saw: his notebooks are clear evidence of this.

'... and take a note with rapid strokes, thus, in a little book that you should always carry with you.'

Vasari commented on how alert the artist was, noting that he was always fascinated when he saw someone of striking appearance. He would follow the person 'all day long', and imprint them so clearly in his mind's eye that when he got home he could draw the person as if they were standing there 'in the flesh'.

'You should go about and often as you go for walks observe... the actions of the men themselves and of the bystanders.'

His friends added their own stories of his working methods. Giampaolo Lomazzo recorded this story: 'Leonardo once wished to make a picture of some laughing peasants. He chose certain men whom he thought appropriate for his purpose, and, after getting acquainted with them arranged a feast for them with some friends. Sitting close to them he then proceeded to tell the maddest and most ridiculous tales imaginable, making them laugh uproariously. Whereupon he observed all their gestures very attentively, and impressed them on his mind; and after they had left, he retired to his room and there made a perfect drawing which moved those who looked at it to laughter as if they had been moved by Leonardo's stories at the feast.'

'... consider the circumstances and behaviour of men as they talk and quarrel, or laugh or come to blows with one another.'

Judging your own pictures

THERE IS A lovely story of Leonardo returning time and again to stand for hours before his fresco of *The Last Supper*, adding only a few strokes of paint at any one time.

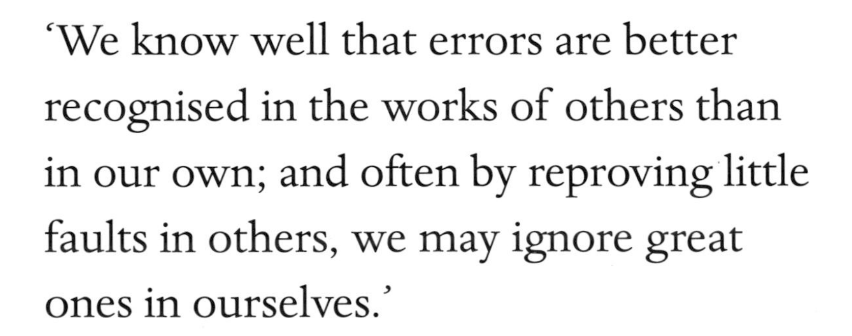

'We know well that errors are better recognised in the works of others than in our own; and often by reproving little faults in others, we may ignore great ones in ourselves.'

'Sitting too close at work may greatly deceive you.'

'It is well that you should often leave off work and take a little relaxation, because when you come back to it you are a better judge.'

'It is good to retire a distance because the work looks smaller and your eye takes in more of it at a glance and sees more easily the lack of harmony and proportion in the limbs and colours of the objects.'

Representing men

LEONARDO WAS a keen observer of human nature, watching his fellow man in all his peculiarity. He searched carefully for the characters to people his paintings.

'Pay attention to them in the streets and piazze and fields, and note them down with a brief indication of forms; thus for a head make an O, and for arm a straight or a bent line, and the same for the legs and body, and when you return home work out these notes in a complete form.'

Leonardo certainly took his own advice when he came to create the characters of Christ's disciples in *The Last Supper:*

'One who was drinking and has left the glass in its position and turned his head towards the speaker. Another, twisting the fingers of his hands, turns with stern brows to his companions.'

'Another with his hands spread shows the palms and shrugs his shoulders up to his neighbour's ear and he as he listens, turns towards him to lend an ear while holding a knife in one hand, and in the other the loaf half cut through.'

Stimulating the imagination

DESPITE HIS concern about the 'artificial' creation of harmonious proportion in the composition of his paintings, Leonardo recommended that the artist should return continually for stimulation to forms in nature, however those forms might be manifest.

'... you can see various battles, and lively postures of strange figures, expressions on faces, costumes and an infinity variety of things, which you can reduce to a good integrated form. Look into the stains of walls, or ashes of a fire, or clouds, or mud or like places, in which, if you consider them well you may find really marvellous ideas.'

'Look at walls splashed with a number of stains, or stones of various mixed colours. If you have to invent some scene, you can see there resemblances to a number of landscapes... mountains, rivers, rocks, trees, great plains, valleys hills, in various ways.'

'The mind of the painter is stimulated to new discoveries, the composition of battles of animals and men, various compositions of landscapes and monstrous things... by indistinct things the mind is stimulated to new inventions.'

How to represent night

THE COMPOSITION of Leonardo's paintings was one of the most important things that the artist was called upon to do. Leonardo gave his students some helpful advice about how to tackle large-scale compositions of difficult subjects. Night offered an obvious problem.

'That which is entirely bereft of light is all darkness. Arrange to introduce a great fire. Then the thing which is nearest the fire will be the most tinged with its colour... those who stand by the sides should be half dark and half in ruddy light. Those visible beyond the edges of the flames will be all lit up by the ruddy glow against a black background.'

Opposite: *St John the Baptist*

The way to represent a battle

SADLY LOST to us, apart from a few sketches, are Leonardo's fresco battle scenes. However, paintings of this kind are not omitted from his advice to artists: in fact those that he describes in greatest detail are paintings of subjects rooted in conflict. Leonardo's composition for *The Battle of Anghiari* fresco was full of violent and naturalistic movement; unfortunately the fresco suffered an all too common fate as Leonardo's experimentation in materials once again failed. Descriptions still survive in his notebooks:

'Represent first the smoke of the artillery, mingled in the air with the dust tossed up by the movement of horses and combatants.'

'You will make the conquerors rushing onwards with their hair and other light things streaming in the wind, with brows bent down.'

'Make a horse dragging the dead body of his master, and leaving behind in the dust and mud the track where the body was dragged along.'

'Make the conquered and beaten pale, with brows raised and knit, and the skin above their brows furrowed with the pain... the lips arched displaying the upper teeth, and the teeth apart as with the crying of lamentation.'

How to represent a tempest

THE STUDY of nature made Leonardo acutely aware that, despite the harmony at the heart of it, it is also supremely destructive.

'In order to represent this tempest you must first show the clouds riven and torn and flying with the wind, together with storms of sand blown up from the sea-shores, and boughs and leaves swept by the strength and fury of the gale...'

'The very air should strike terror through the deep darkness caused by the dust and mist and heavy clouds.'

'Make the clouds driven by the impetuous winds, hurled against the high mountain tops, and there wreathing and eddying like waves that beat upon the rocks.'

'Of the men who are there, some should have fallen and be lying wrapped round by their garments, and almost indistinguishable on account of the dust, while those who remain standing should be behind some tree with their arms thrown round it that the wind may not tear them away.'

Representing a deluge

'People on trees that cannot support them, trees and rocks, towers and hills crowded with people, hills covered with men, women and animals, with lightning from the clouds which illuminates the scene.'

WHAT IS included in these images is in effect a summary of all that Leonardo wrote of the destructive power of nature. This is not nature as the giver and sustainer of life, but nature as destroyer, before which man stands powerless.

'Darkness, wind, tempest at sea, deluge of water, forests on fire, rain, bolts from heaven, earthquakes and destruction of mountains, levelling of cities.'

'Whirlwinds which carry water and branches of trees, and men through the air. Branches torn away by the winds crashing together at the meeting of the winds with people upon them.'

LEONARDO'S SCIENTIFIC VISION

VASARI WROTE, 'He used to make models and plans showing how to excavate and tunnel through mountains without difficulty, so as to pass from one level to another; and he demonstrated how to lift and draw great weights by means of levers, hoists and winches, and ways of cleansing harbours and using pumps to suck up water from great depths. His brain was always busy on such devices.'

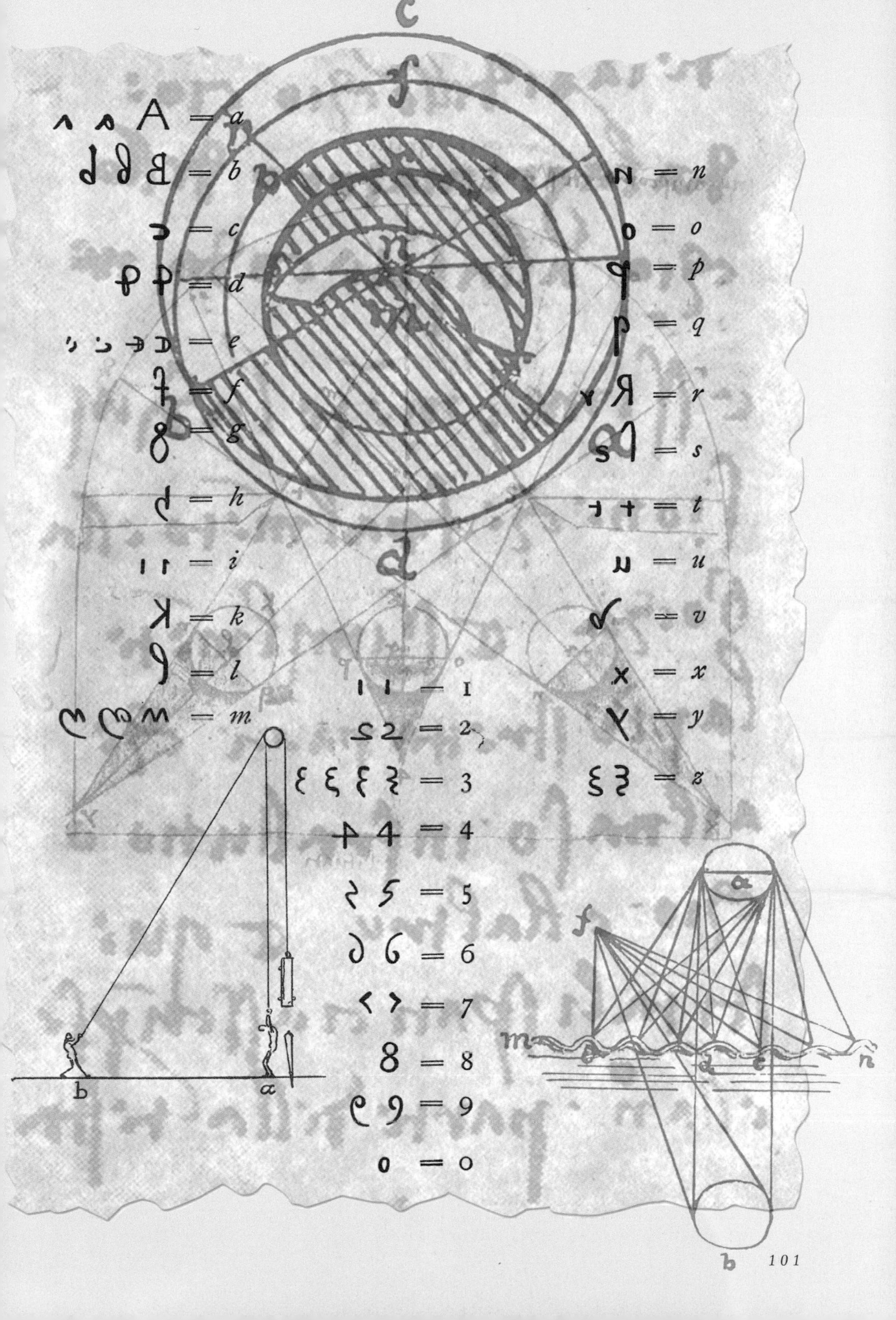

= *a*
= *b*
= *c*
= *d*
= *e*
= *f*
= *g*
= *h*
= *i*
= *k*
= *l*
= *m*
= *n*
= *o*
= *p*
= *q*
= *r*
= *s*
= *t*
= *u*
= *v*
= *x*
= *y*
= *z*

= 1
= 2
= 3
= 4
= 5
= 6
= 7
= 8
= 9
= 0

Leonardo's mind overflowed with invention and he did more, perhaps than anybody else in history, to lay down our modern approach to science. He had always studied nature but in the later part of his life, however, Leonardo's primary interest became scientific investigation. Two of his major areas of interest, anatomy and the study of fluids, far exceeded anything that had been achieved before him.

The range of his mind was breathtaking: anatomy, zoology, botany, geology, optics, aerodynamics and hydrodynamics, to name only some of his interests. He was consumed by curiosity. Although he consulted Classical authorities, he was a trained artist rather than an academic, and therefore his grounding was biased towards observation, and he recorded what he saw. To offer that gift to the service of science enabled him to produce startling developments in a vast range of scientific study and, in the process, he also produced some of the most beautiful drawings ever created.

Yet, despite Leonardo's vast curiosity and inextinguishable energy, in moments of depression he felt he was bound to fail, comparing himself with those he most vehemently dismissed: the alchemists and other pursuers of false science.

'I wish to work miracles; it may be that I shall possess less than other men of peaceful lives, or those who want to grow rich in a day. I may live for a long time in great poverty, as always happens and to all eternity will happen, to alchemists, the would-be creators of gold and silver, and to engineers who would have dead water stir itself into life and perpetual motion, and to those supreme fools, the necromancer and the enchanter.'

Leonardo's thinking was so far ahead of his times he was bound to be the focus of jealousy, misunderstanding and suspicion. In a time when autopsies were viewed with extreme suspicion by the Church, Leonardo was driven by a determination to discover the fundamental truths he was prepared to suffer for the sake of science. As Paolo Giovi noted: 'In the medical faculty he learned to dissect the cadavers of criminals under inhuman, disgusting conditions... because he wanted [to examine and] to draw the different deflections and reflections of limbs and their dependence upon the nerves and the joints.'

Whilst laying down much of the foundation of our own way of thinking and proceeding in the modern world, Leonardo was only too aware of the imperfection and incompleteness of his knowledge. It was this awareness that made him complain of his inability to finish anything. He had all kinds of plans to publish the result of his researches, but his fear of exposure and his acknowledgement of the incomplete state of his understanding prevented him from disclosing his discoveries. Vasari commented, 'Leonardo's profound and discerning mind was so ambitious that this was itself an impediment; and the reason he failed was because he endeavoured to add excellence to excellence and perfection to perfection. As our Petrach has said, the desire outran the performance.'

For this reason his work remained either unknown or inaccessible until the world had caught up with, or often surpassed, his thinking.

Leonardo was known for his unorthodox views and his unwillingness to accept scriptural authority. For instance, he rejected the theory of the biblical Flood being responsible for depositing fossils many miles from their origin and, instead, deduced from his studies of rock formations that they represented long spans of geological time.

Leonardo was desperate to discover the truth, regardless of accepted opinion and regardless of personal cost. Even when everything seemed hopeless, he was still driven to perform tasks that he suspected, despite his best efforts, would end in frustration.

Leonardo had formulated for himself a system based on observation, analysis and experiment. Taking Euclid as his model he understood that the laws of nature were capable of mathematical expression. He appreciated that with the aid of scientific investigation it was possible to pursue effects back to their cause, and that man was capable of harnessing, through his understanding of the laws of nature, the powers that nature possessed.

'Experience does not feed investigators on dreams, but always proceeds from accurately determined first principles, step by step in true sequence to the end.'

The empirical approach

BEING APPRENTICED to Andrea del Verrochio perfectly suited Leonardo. Apparently he was delighted to be in Verrochio's workshop, and rapidly became a first-class geometrician.

His training as an artisan meant that Leonardo knew no Latin and so could not read Classical texts. It was not until his thirties that he decided to learn Latin on his own, but whether he achieved success is doubtful: his notebooks are written in robust Italian.

In practical terms, this lack of Greek and Latin meant that rather than relying on Classical teachings he was determined to look at subjects afresh.

'All our knowledge has its origins in our perceptions.'

His appreciation of the world was rooted in observation, a habit that served him equally as an artist and in his study of science.

'To me it seems that all sciences are vain and full of errors that are not born of Experience... that is to say, that do not at their origin, middle or end, pass through any of the five senses... where reason is not, its place is taken by clamour. This never occurs where things are certain. Therefore, where there are quarrels, there true science is not... wherever it is known controversy is silenced for all time.'

Demonstration

MATHEMATICS was central to Leonardo's thinking. Taking his lead from Euclid's geometrical proofs, his mathematical demonstration followed a logical process.

'There is no certainty where one can neither apply any of the mathematical sciences nor any of those which are connected with the mathematical sciences.'

Although much of his understanding was intuitive, Leonardo understood completely the need for analytical thought.

'Here you must proceed methodically; that is you must distinguish between the various parts of the proposition so that there may be no confusion and you may be well understood.'

'See to it that the examples and proofs that are given in this work are defined before you cite them.'

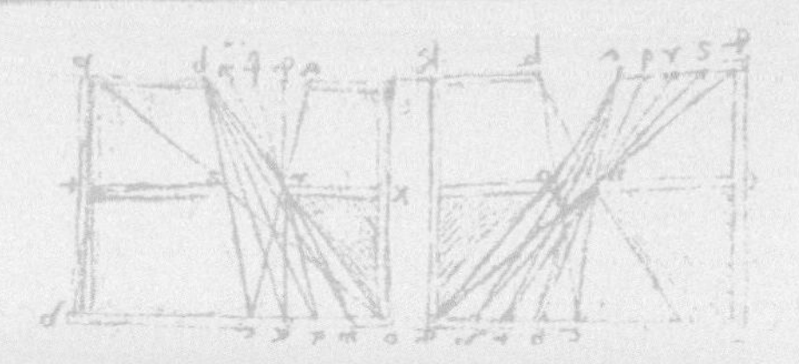

'Those who condemn the supreme certainty of mathematics feed on confusion, and can never silence the contradictions of the sophistical sciences which lead to eternal quackery.'

Experiment

THE SCIENTIFIC process, defined by Leonardo, was concluded with repeated experimentation to test the validity of the conclusions drawn. Only after a rigorous process of this kind would it be possible to trust those conclusions and proceed to employ them in the act of inventions.

'Before you base a law on this case test it two or three times and see whether the tests produce the same effects.'

What Leonardo also understood was that by close observation and reflection, new ideas inevitably arose. And in the process, he was getting closer to the 'first cause', the origin of the creative process.

Opposite: *Archemedes' screw*

'This experiment should be made many times so that no accident may occur to hinder or falsify this proof, for the experiment may be false whether it deceived the investigator or not.'

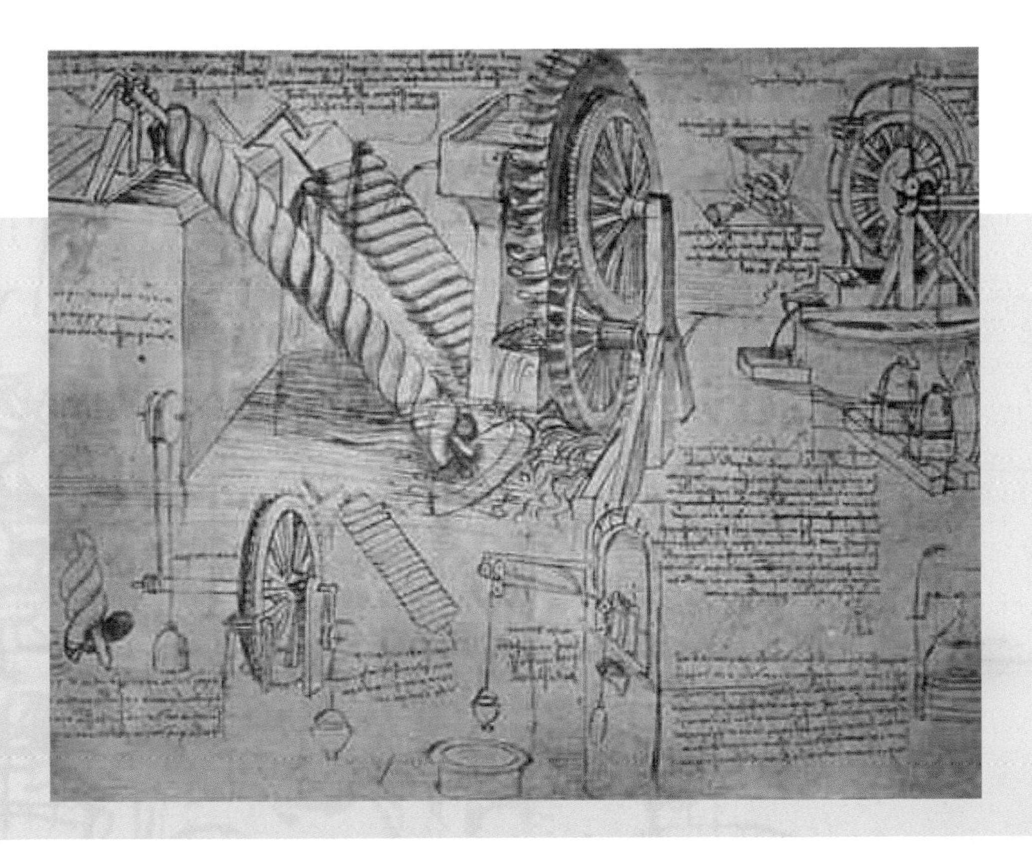

'Mechanics is the paradise of mathematical science, because by means of it one comes to the fruits of mathematics.'

The four elements

THE ACCEPTED view in Leonardo's time, handed down from the Classical thinkers, was that an omnipotent creator had made the physical world in the form of a series of concentric spheres, at the centre of which was the earth surrounded by spheres of water, air and fire, with ether encompassing it all.

Despite this being the accepted theory, there is no doubt that Leonardo attempted to test its validity by empirical methods. He was particularly fascinated by water. He thought that this was the fundamental driving force in nature, and studied it in all its forms throughout his life.

'Every part of the depth of the earth in a given space is composed of layers, and each layer is composed of heavier and lighter parts; the lowest being the heaviest. And the reason for this is that these layers are formed by the sediment from the water discharged into the sea by the current of the rivers which flow into it...These layers of soil are visible in the banks of rivers which in their continuous course have cut through and divided one hill from another in a deep defile, wherein the waters have receded from the shingle of the banks; and this has caused the substance to become dry and to turn into hard stone.'

Water and earth

From his own observation of the landscape and his studies of the ancient philosophers, Leonardo noted that the earth, although the most dense of elements, was continually affected by the other elements, particularly water.

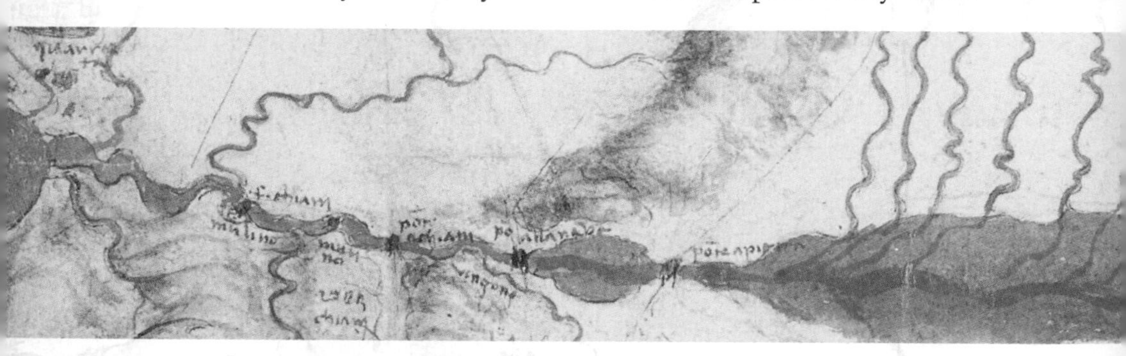

'... against the irreparable inundation caused by swollen and proud rivers no resource of human foresight can avail; for in a succession of raging and seething waves gnawing and tearing away high banks, growing turbid with the earth from ploughed fields... and uprooting the tall trees, it carries these as its prey down to the sea which is its lair.'

As ever, his conclusions were drawn from first-hand observation. Here for instance is a description of flooding that he witnessed whilst travelling in Savoy.

'... certain forests sank in and left a very deep gap; and about four miles from there the earth opened like a gulf in the mountain, and through a sudden immense flood of water which scoured the whole little valley of tilled soil, vineyards, and houses, and wrought the greatest damage wherever it overflowed.'

The qualities of water

WE CAN IMAGINE Leonardo walking by riverbanks, studying the flow of mountain streams, or the waves of the ocean, or the perturbation of water as it flowed along canals.

'Of the four elements water is the second in weight and the second in respect of mobility. It is never at rest until it unites with the sea...'

He carried out numerous experiments to attempt to understand the nature of water, the pressure it exerted and the power it generated.

'It readily raises itself by heat in thin vapour through the air. Cold causes it to freeze. Stagnation makes it foul. That is, heat sets it in motion, cold causes it to freeze; immobility corrupts it.'

'It is the expansion and humour of all vital bodies. Without it nothing retains its form. By its inflow it unites and augments bodies.'

'It assumes every odour, colour and flavour and of itself it has nothing.'

'It is now sharp, now strong, now acid and now bitter, now sweet and now thick or thin, now it seen bringing damage or pestilence and then health, or, again, poison.'

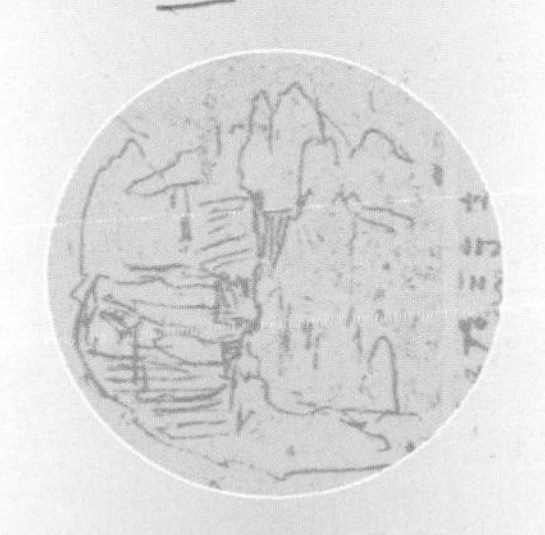

Water in motion

UNDERSTANDING and controlling the element of water was seen as key, as water was a crucial means of irrigation, transport and power in Leonardo's day.

Nobody before him had set out to study the properties of water in the systematic way that Leonardo did, nor come to such a complete comprehension of its nature.

'Water is the driver of nature.'

'If you throw a stone in a pond... the waves which strike against the shores are thrown back towards the spot where the stone struck; and on meeting other waves they never intercept each other's course... In a small pond one and the same stroke gives birth to many motions of advance and recoil.'

'The greater wave is covered with innumerable other waves which move in other directions; and these are deep and shallow according to the power that generated them…'

'A wave of the sea always breaks in front of its base, and portion of the crest will then be lower which before was highest.'

Air

As a painter, Leonardo observed the consistency of air: air suffused with energy in the form of sunlight; with water in the form of clouds, mist and rain; and with earth in form of dust and smoke. He was fascinated both by the movement of air and the movement of objects through air. He studied wind and weather, aerodynamics and acoustics.

'The cloud or vapour that is in the wind is produced by heat and is smitten and banished by the cold, which drives it before it, and where it has been ousted the warmth is left cold. And because the cloud which is driven cannot rise because the cold presses it down and cannot descend because the heat raises it, it therefore must proceed across; and I consider it has no movement of itself, for as the said powers are equal they confine the substance that is between them equally.'

'Fitful impetuosity of the wind is shown by the dust that it raises in the air in its various twists and turns... one sees how the flags of ships flutter in different currents; how on the sea one part of the water is struck and not another; and the same thing happens in the piazze and on the sandbanks of rivers, where dust is swept away furiously in one part and not in another.'

As an accomplished musician he was intrigued by the way sound travelled.

'Just as the stone thrown into the water becomes the centre and cause of various circles, [so] the sound made in the air spreads out in circles and fills the surrounding parts with an infinite number of images of itself.'

The colour of atmosphere

LEONARDO was deeply fascinated by air saturated by pure colour.

'I say that the blue which is seen in the atmosphere is not its own colour but is caused by warm humidity evaporated in minute and imperceptible atoms on which the solar rays fall rendering them luminous against the immense darkness of the region of fire that forms a covering above them.'

Like a physical symbol of all his investigations into the laws that governed creation, Leonardo also wrote about wanting to transcend the atmosphere and peer into the 'sphere of the element of fire'.

Opposite: *The Annunciation*

He saw that the interaction of the light of the sun and air saturated with tiny particles not only created the blue of the sky but also all the other colours which he recreated in his paintings.

'...the atmosphere assumes this azure hue by reason of the particles of moisture which catches the luminous rays of sun.'

The sun

THE DESIRE to study the 'sphere of fire' led Leonardo to use his ingenuity to peer out into the vastness of the heavens. Up in his roof, under a skylight, it is believed that he used some kind of optical instrument: he left a 'to do' note for himself: 'Construct glasses to see the moon magnified'. He speculated on the movement of the sun in relation to the earth and appeared to be of the same mind as Copernicus, who held that the earth moved round the sun.

'In the whole universe I do not see a body of greater magnitude and power than this, and its light illumines all celestial bodied which are distributed throughout the universe.'

Leonardo acknowledged that the power of the sun drove everything, and like many thinkers of his time identified the enlivening power of the sun with the vital spark in men:

'All vital force descends from the sun since the heat that is in living creatures comes from the soul (vital spark); and there is no other heat nor light in the universe.'

'The sun does not move.'

'The motion of the elements arises from the sun.'

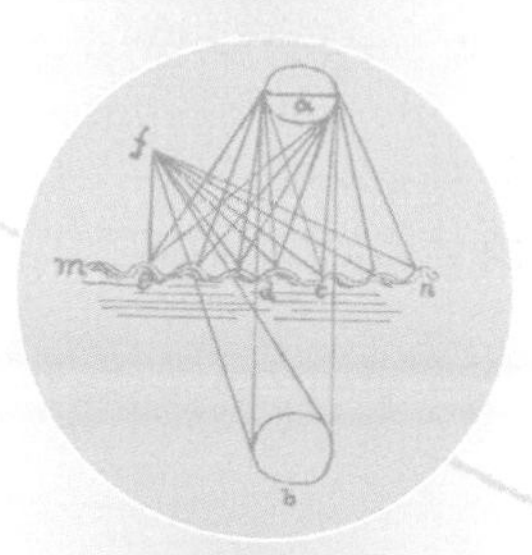

'The light and heat of the universe comes from the sun, and its cold and darkness from the withdrawal of the sun.'

Nature's laws

As a lover of nature and as a creative thinker, Leonardo was certain that mankind's destined task, born in the image of the Creator, was to come to a direct understanding of the laws of nature. In a practical sense he seemed to understand, as few had done before him, the need for a total marriage of theory and practice. He knew that there could be no major developments in technology without an understanding of how 'Nature' herself created.

'Nature begins with the cause and ends with the experience; we must follow the opposite course, namely, begin with the experience, and by means of it investigate the cause.'

 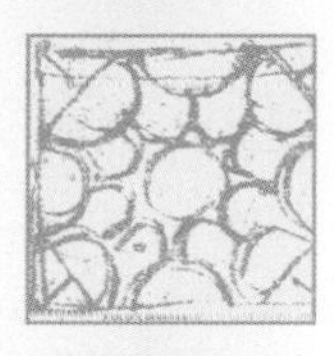 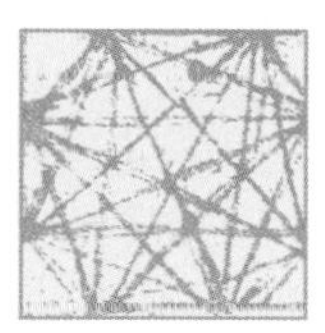

'Nature does not break her law; Nature is constrained by the logical necessity of her law which is inherent in her.'

'Necessity is the theme and inventor of nature, its eternal law and curb.'

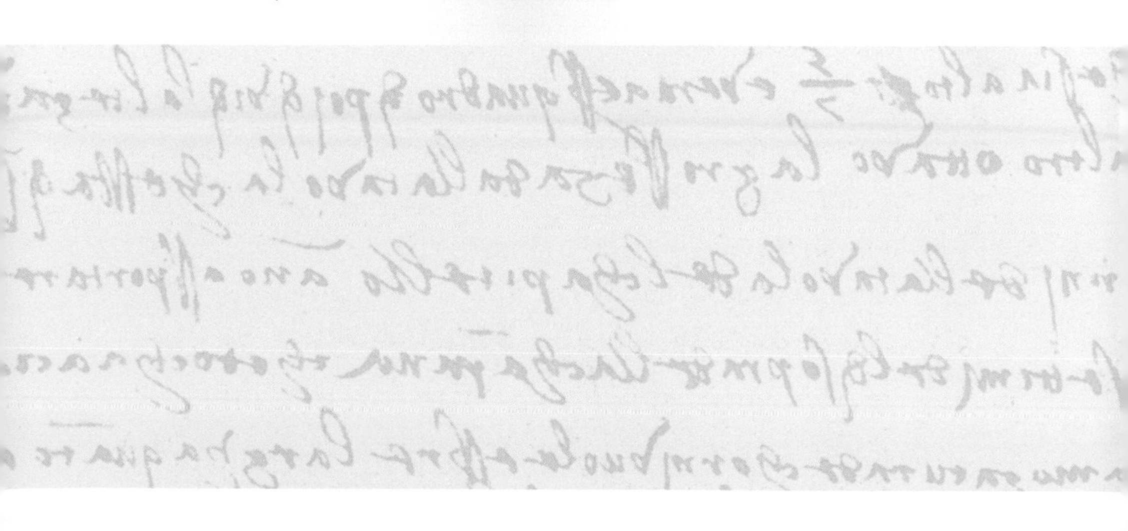

The powers of nature

IN THE INTERACTION of the elements, Leonardo saw the powers of nature at work: weight, force, movement and percussion. These forces were everywhere: in the heat of the sun, in the movement of water, in growth and form.

An artist craftsman was inevitably asked how to solve practical problems, and Leonardo was a genius at tackling practical problems. It is not entirely clear when he started asking himself 'why'. But at some stage, the natural philosopher in him realized the importance of discovering and working from the fundamental laws of nature:

'First deal with weight, then with movement, which creates force, then with this force and lastly with the blow.'

In other words, all Leonardo's study of the laws of nature, and the vast flow of inventiveness that followed from it, were born from the desire to know. More fundamental questions appear in his Milan notebooks: in these he is seen moving away from Florence and his artisan background, and beginning to search out the cause of things.

One result of this way of thinking was a planned book on theoretical mechanics.

'Weight, force, and impetus are the children of movement because they are born from it. Impetus is frequently the cause why movement prolongs the desire of the thing moved.'

'Weight and force together with the motion of bodies and percussion are the four powers of nature by which the human race in its marvellous and various works seems to create a second nature in this world, for by the use of such powers all the visible works of mortals have their being and their death.'

Weight

LEONARDO HAD a number of theories about the nature of gravity, one being that the impression of weight and lightness only occurred in the relationship of one element to another, and that the elements in themselves possessed no actual weight.

'Every gravity weighs through the central line of the universe because it is drawn to this centre from all parts.'

'Weight is caused by one element being situated in another; and it moves by the shortest line towards its centre, not by its own choice, not because the centre draws it to itself, but because the other intervening element cannot withstand it.'

'Heat and cold proceed from the nearness or remoteness of the sun. Heat and cold cause the movement of the elements. No element has of itself weight or lightness.'

Force

LEONARDO'S THOUGHTS on the nature of force are fascinating. He saw it as a spiritual energy, an invisible power that animated the inanimate for a short while, giving it an appearance of life:

'It is born in violence and dies in liberty; and the greater it is the more quickly it is consumed.'

'Force is spiritual essence which by accidental violence is united to weighty bodies, restraining them from following their natural inclination...'

'It drives away in fury whatever opposes its destruction. It desires to conquer and slay the cause of its opposition, and in conquering destroys itself.'

'Without force nothing moves.'

Movement

LEONARDO BELIEVED that the spiritual energy of force created movement. The point of equilibrium was disturbed and movement took place until the energy that impelled the movement was exhausted.

'No movement can ever be so slow that a moment of stability is found in it.'

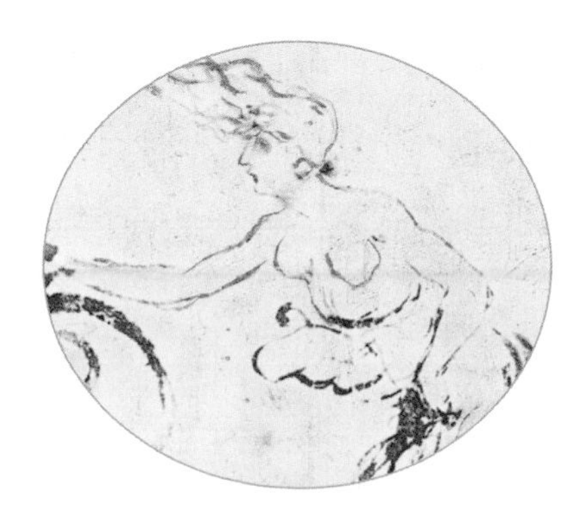

'Some say that the arrow in moving propels a wave of air in front of itself, and that this wave by means of its movement prevents the course of the arrow being impeded. This is incorrect... the air which passes in waves in front of thc arrow does so because of the movement of this arrow, and it lends little or no help of movement to its mover.'

Contained in his notebooks, in addition to observations about movement in spacc, are observations about time.

'A point has no part; a line is the transit of a point; points are the boundaries of a line. An instant has no time. Time is made of movements of the instant, and instants are the boundaries of time.'

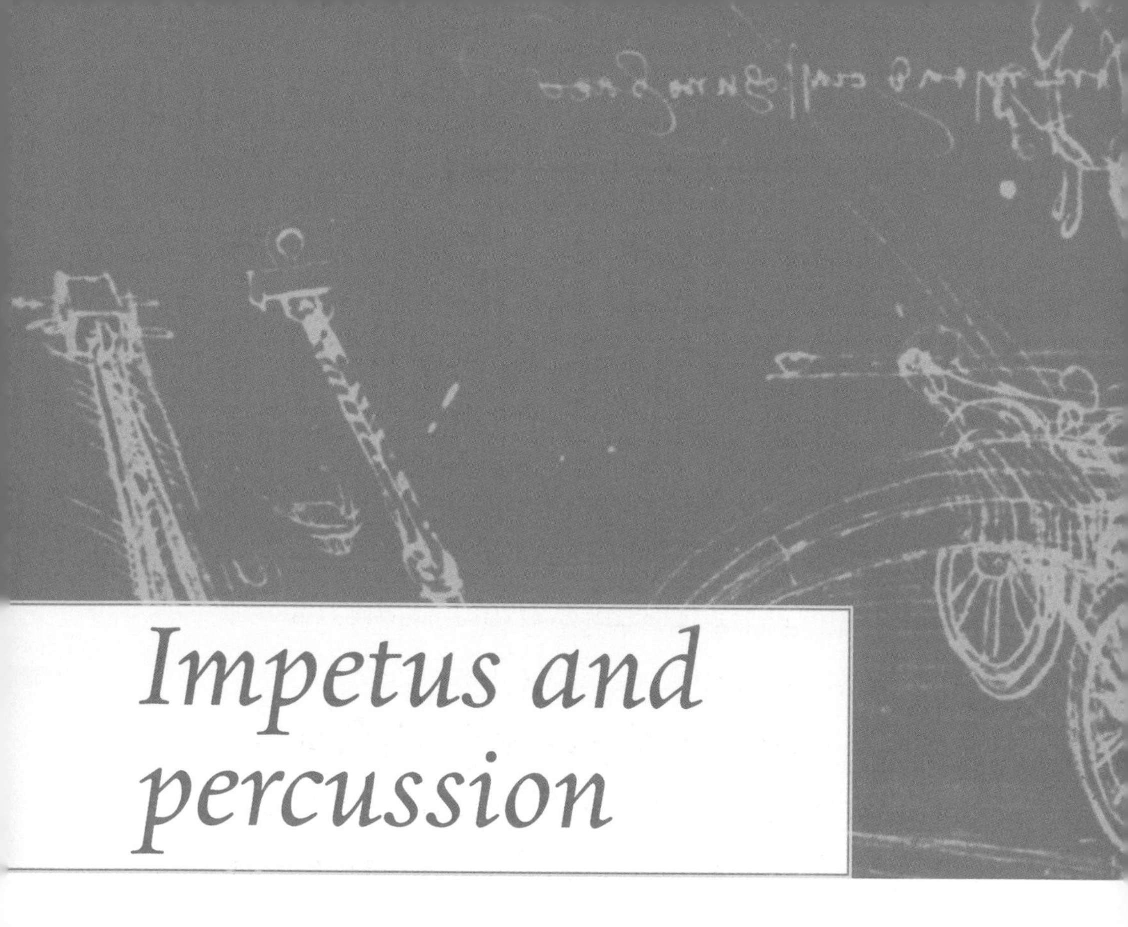

Impetus and percussion

LEONARDO'S intention was to study the universal laws of dynamics. Eventually, however, Leonardo found more satisfaction in the practical application of these laws – he was, after all, a creative artist.

'At the beginning of movement there is impetus, which is resisted from its very conception. Worn down by friction it exhausts itself and finally achieves rest. Alternatively it is resisted more violently resulting in percussion.'

'There are two different kinds of percussion, simple and complex. The simple is made by the movable thing in its falling movement upon its object. Complex is the name given when this first percussion passes beyond the resistance of the object which it first strikes, as in the blow of the sculptor's chisel which is afterwards transferred to the marble that he is carving.'

'... percussion is also the cause of sounds and the breaker and transmuter of various things and the product of a second motion.'

The scientific programme

THE SCIENTIFIC procedure Leonardo pursued was a four-step process:

1. Sensory observation is the beginning of scientific theory.

2. Reason then reflects upon these observations and deduces from them scientific laws.

3. These laws are then logically demonstrated like mathematical propositions.

4. Finally the experiment is tested.

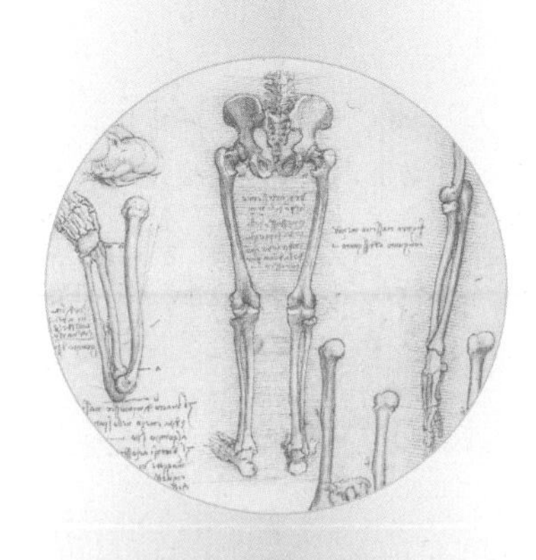

After this process had been completed the intention was to apply what had been learned to the creation of new inventions.

'What trust can we place in the ancients, who tried to define what the Soul and Life are – which are beyond proof – whereas those things which can at any time be clearly known and proved by experience remained for many centuries unknown or falsely understood.'

'The senses are of the earth. Reason stands apart from them in contemplation.'

'Experience (is) the interpreter between formative nature and the human species...'

Inventions

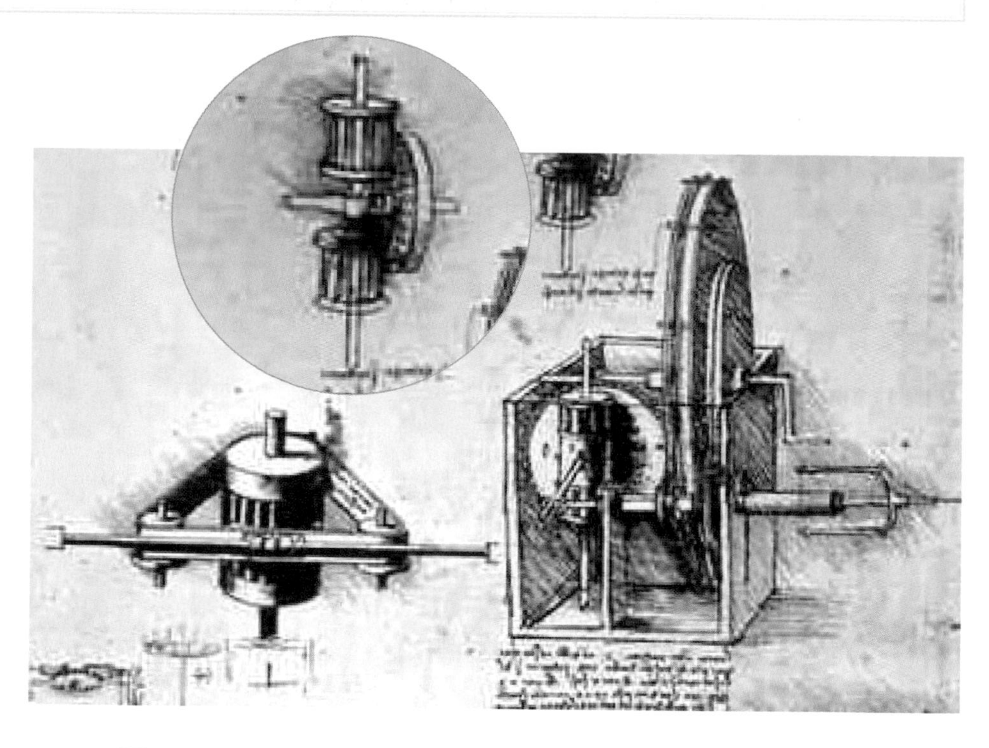

IN LEONARDO'S notebooks we see evidence of plans for a book on the elements of machines, in which he looks not so much at individual inventions but those elements of mechanical engineering which would enable him to solve any problem. He mapped out what he planned to say:

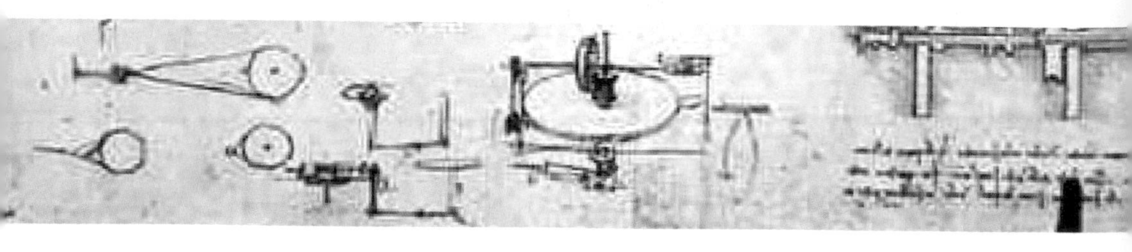

'We shall briefly discuss the many ways there are to use screws, the various types of endless screws, and the many motions that are performed without screws... We shall also speak of inverted screws and screws that by a single motion thrust and pull a weight... Account too, will be given to their nature and uses, their composition, levers and utility.'

'We shall also deal with the differences existing between a lever operating with constant force, that is the wheel, and the lever of unequal force, that is, the straight beam, and why the former is better than the latter and the latter more compact and convenient than the former. We shall also discuss the ratchet wheel and its pawl, the flywheel and the impetus of its motion...We shall describe how air can be forced under water to lift very heavy weights.'

Machines

LEONARDO devised not only machines of destruction but also machines for propulsion, including an early bicycle. He designed, among many other things, windlasses, machines for making files, for carding wool, and for drilling and reaming, as well as several different types of loom. Leonardo's inventions prefigured many of the machines that centuries later would promote the Industrial Revolution.

'If the spring is of uniform thickness, its power diminishes gradually as it is unwound.'

'When you make a screw that engages only a single tooth on the wheel, it will be necessary to add a pawl in order to avoid the reversal of the wheel's motion should the tooth break.'

'This lifting device has an endless screw which engages many teeth on the wheel. For this reason the device is very reliable.'

'It is customary to oppose the violent motion of the wheels of a clock driven by their counterweights driven by certain devices called escapements, as they keep the timing of the wheels which move it. They regulate the motion according to the required slowness and the length of the hours.'

The inventions of war

IT WAS IN the realm of munitions that Leonardo left his largest legacy of mechanical inventions. He lived at a time when things were moving rapidly in the technology of warfare: the time was ripe for the development of devices that could deliver more powerful shot more rapidly.

'I have also plans of mortars most convenient and easy to carry with which hurl small stones in the manner almost of a storm.'

In his inventions for attack and defence are designs for all kinds of ballista (catapult), cannon and aquebus (musket). For these Leonardo developed a variety of technological refinements, including rapid-fire mechanisms, which were the forerunners of the modern-day machine gun.

'When a place is besieged, I know how to remove the water from the trenches and to construct an infinite number of bridges, covered ways and ladders...'

'I have plans for destroying every fortress or other stronghold even if it were founded on a rock.'

One of his notebooks features the famous drawing of a huge ballista and a mortar driven by steam, which was much like the cannon that was finally used in the nineteenth century. There was also amongst these inventions the precursor of the tank.

'I will make covered cars, safe and unassailable, which will enter among the enemy with their artillery, and there is no company of men at arms so great that thcy not break it.'

Although these designs were the devotion of his later years, at the age of thirty Leonardo listed an impressive array of designs of weaponry that he was able to offer. Although at this stage there is little in the notebooks to substantiate his claims, it wasn't entirely bravura. Already he had mastered ratchets and screws, levers and cogs – all the basic ingredients of mechanised warfare.

Wings and flight

EXAMINING MOTION through the air, Leonardo made a thorough study of the flight of birds and how their physical structure enable them to resist the force of gravity and fly. Above all he studied the structure of wings:

> 'The wing of a bird is always concave in its lower part extending from the elbow to the shoulder, and the rest is convex. In the concave part of the wing the air is whirled round, and the convex is pressed and condensed.'

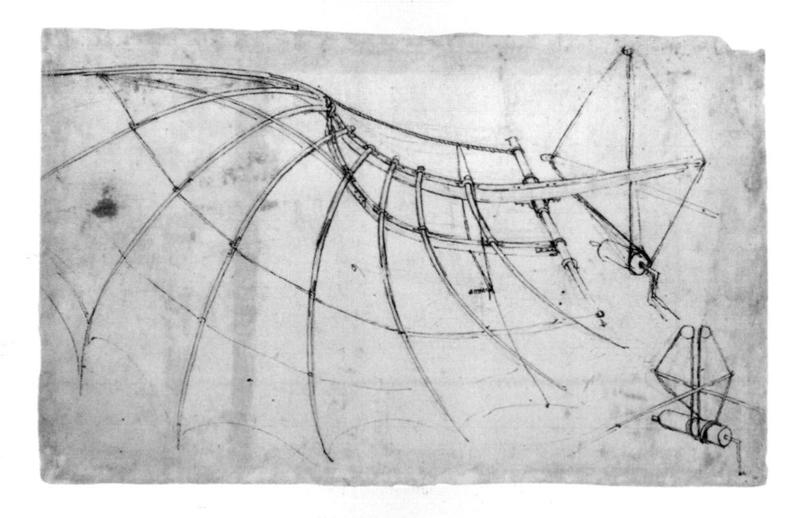

Opposite: *Leda and the swan*

'To speak of this subject you must... explain the nature of the resistance of the air, in the second the anatomy of the bird and its wings, in the third the method of working the wings in their various movements, in the fourth the power of the wings and the tail when the wings are not being moved and when the wind is favourable to serve as guide in various movements.'

'The sinews beneath the birds' wings are more powerful than those above. The shoulder, being the helm of the wing, is hollow below like a spoon; and being concave below, it is convex above. It is fashioned thus that the process of going up may be easy, and that of going down difficult and meeting with resistance; and it is especially adapted for going forward drawing itself back in the manner of a file.'

'Bats when they fly must of necessity have their wing completely covered with a membrane, because the creatures whereon they feed seek to escape by means of... various twists and turns.'

Movement through air

LEONARDO'S STUDY of flight is directly related to his study of the elements. In preparing a book on the mechanics of flight, he wanted to come to an understanding of the movement of the air and how things passed through it.

He also asked looked at how mechanical movement could take place within air, how things heavier than air could keep aloft and how he could create a flying machine.

'The air moves like a river and carries the clouds with it; just like running water carries all things that float upon it.'

'The movement of the air against a fixed thing is as great as the movement of the moving thing against the air that does not move.'

And of the science of the weight proportioned to the powers of their movers:

'The force of the mover ought always to be proportionate to the weight of the thing to bc moved and to the resistance of the medium in which the weight moves.'

'...in the flight of the birds, the sound that they make with their wings in beating the air is deeper or shriller according to whether the movement of the wings is slower or swifter.'

Flying machines

FROM CHILDHOOD, Leonardo was fascinated by the flight of birds. He admired the sense of freedom they possessed; he was known to go to markets and buy up all the caged birds so that he could open the cage doors and thus give them back their lost freedom.

Given his delight in flight, his study of the elements and his genius as an inventor, it was inevitable that Leonardo would want to design a flying machine and dream of taking wing himself.

'The man in the flying machine must be free from the waist upwards in order to balance himself as he in a boat, so that his centre of gravity and that of his machine may counterbalance each other and shift where necessity demands.'

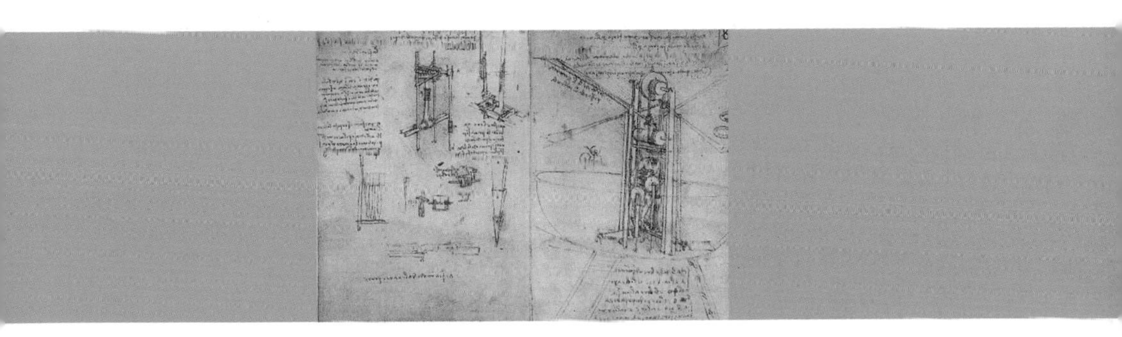

'A bird is an instrument working according to mathematical law, which instrument it is in the capacity of man to reproduce... but not with as much strength, deficient only in the power of maintaining equilibrium. We may therefore say that such an instrument constructed by man is lacking in nothing except the life of the bird.'

LEONARDO AS THINKER

LEONARDO WROTE about many things, among them his *Prophecies*. His prophecies were part riddle and part genuine prophecy. Given Leonardo's consuming interest in the subject, one of them concerns flight:

'Feathers will raise men, as they do birds towards heaven.'

On this occasion he was not talking about physical flight but the flight of the mind. The feathers he referred to were the very quills with which he wrote his notes.

Some of the aphorisms and fables in his notebooks were clearly his private thoughts, but some were likely to have been composed to provide courtly entertainment.

As part of his role as, in his time, various courts' entertainer, Leonardo was expected to design sets and costumes for court masques.These presentations were loaded with imagery and

philosophical significance. Little is now known of the nature of them, but a few descriptions survive:

'The paradise was made to resemble half an egg (displayed hollow side forward) which from the edge inwards was completely covered with gold, with a great number of lights representing stars, with certain gaps where the seven planets were situated, according to their high or low ranks. Around the upper edge of the aforesaid hemisphere were the twelve signs... In the Paradise were many singers, accompanied by many sweet and refined sounds.'

Leonardo was universally recognized as being a master of symbolism and allegory. Whether this was something that he took seriously as part of his duties to his patrons, or it was merely a distraction from his more scholarly studies, is hard to say.

In the notebooks there are drawings of a symbolic nature, mainly related to the courtly entertainments Leonardo was associated with in Milan, and there has been much argument as to how much symbolism Leonardo imparted in his paintings. A well-known early interpretation of *The Virgin and Child with St Anne* was provided by the poet Girolamo Cassio:

St. Anne, as one who knew
That Jesus assumed a human shape
To atone for the Sin of Adam and Eve
Tells her daughter with pious zeal:
Beware if you wish to draw Him back
For the heavens have ordained the sacrifice.

This is an evidently orthodox enough interpretation, but given the heady mix of traditional theology and Neo-platonic and Aristotelian philosophy of Leonardo's day, one would expect him to include symbolic allusions in his paintings. Indeed, Leonardo's biographer, Vasari, described him thus: 'Leonardo was so heretical a cast of mind, that he conformed to no religion whatever accounting it perchance much better to be a philosopher than a Christian.'

However, unlike Botticelli whose most famous works are full of Neo-platonic imagery, there is little to suggest that Leonardo imported into his painting imagery of this kind. This is surprising when one considers that some of his earliest philosophical influences were derived from the Medici court in Florence, which was entirely imbued with Neo-platonic thinking. The Platonic philosophers who influenced all those associated with this court suggested that the arts had a specific function and that was to remind the viewer of the eternal truths which they, by being born into the body, easily forgot. Powerfully presented in pictorial form, these ideas apparently had the ability to remind the viewer of their own essential nature and 'raise the mind to contemplation'.

In addition, what might also be claimed – given the nature and power of his paintings – was that the very thing that this imagery was attempting to achieve, was achieved by Leonardo. For if there is one artist in all of history capable of putting people in touch with their own 'immortal soul', it must be he. Leonardo did this not so much through the use of symbols but by the directness of his vision – his seeming ability to lift the veil of the physical world and to see in depth.

'Painting is the way to learn to know the maker of all marvellous things.'

Leonardo himself claimed that painting was 'a subtle invention, which with philosophy and subtle speculation considers the nature of all forms'. It is also interesting to note that in his attempts to raise the painter from the status of mere artisan, he called painting a 'science': science as it was understood in his own time meant a source of genuine knowledge.

'All our knowledge has its foundation in our senses.'

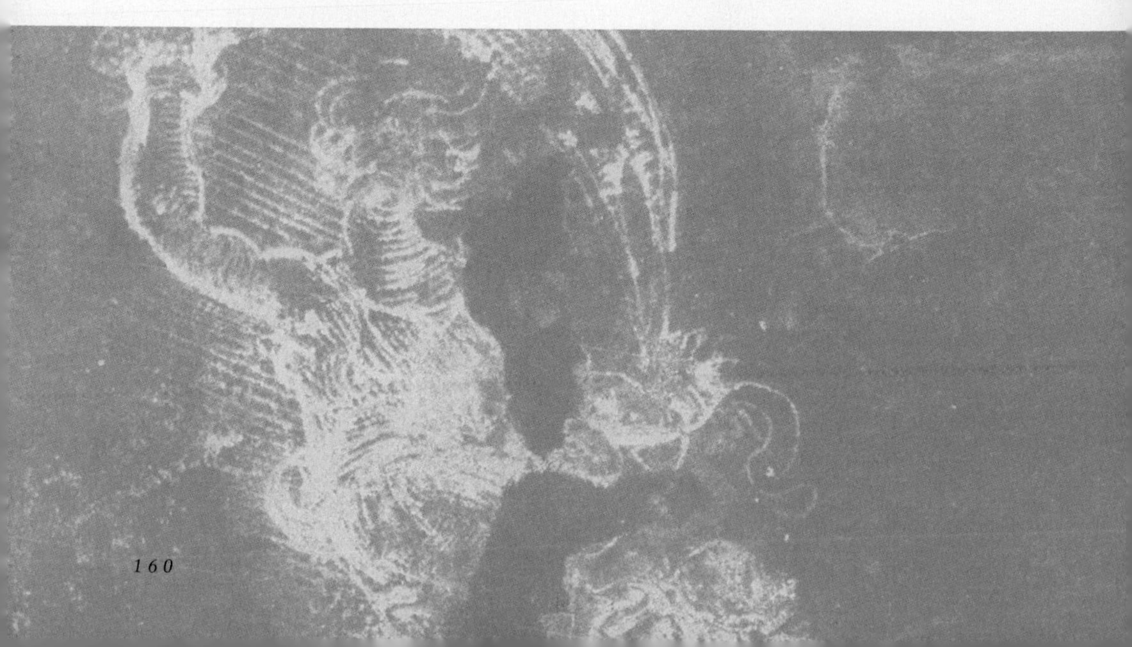

Leonardo did, after all, want to test everything in experience, scornful of those philosophers who indulged in abstract speculation without looking in detail at the knowledge the senses provided. But what he also stated was that there was another step in the scientific process that transcended a completely materialistic approach:

'The senses are earthbound and reason stands outside them when reason contemplates.'

Reason for him was not merely the power to think things through. Reason was the highest faculty of mind, which in the process of understanding had to 'stand outside' in contemplation. Leonardo believed that, in order to understand the laws of nature and to use those laws in being a creator in his own right, mankind had to disengage from the physical world, stand back and reflect.

Comparison of the arts

LEONARDO sang his own songs accompanying himself on a lyre he had made in the shape of a horse's skull. He knew about music at first hand as a composer and performer, but when he came to compare music with painting there was no doubt where his heart lay.

'Music may be called the sister of painting, for she is dependent upon hearing, the sense which comes second and her harmony is composed of the union of proportional parts sounded simultaneously, rising and falling in one or more harmonic rhythms.'

'Painting does not have the need for interpreters for different languages as does literature.'

'By means of her basic principle, that is design, she [art] teaches the architect to make his edifice so that it will be agreeable to the eye, and teaches the composer of variously shaped vases, as well as the goldsmiths, weavers and embroiderers. She has discovered the characters by which different languages are expressed, has given numerals to the arithmetician, has taught us to represent the figures of geometry; she teaches matters of perspective and astronomy, machinists and engineers.'

The power of art

MUCH HAS been written about the power and purpose of art. Leonardo reflected deeply upon what he was attempting to achieve through his painting. One aim was to raise the standing of the artist. True artists, in his estimation, were men of insight and understanding and not only did he convey this vision through his own paintings, he also wrote about it.

'Oh wonderful science that can preserve the transient beauty of mortals and endow it with a permanence greater than the works of nature, for these are subject to the continual change of time which leads them to inevitable old age!'

Paintings produced in this spirit are not of frozen moments in time, but in their unchanging nature they portray unchanging beauty and eternal truth.

'What is fair in men passes away, but not so in art.'

'Truly painting is a science, the true-born child of nature, for painting is born of nature, but to be more correct we should call it the grandchild of nature; since all visible things were brought forth by nature and these her children have given birth to painting. Therefore we may justly speak of it as the grandchild of nature and as related to God.'

Harmony

HARMONY IS achieved when things in all their apparent diversity are drawn together as one.

'Do you know that our soul is composed of harmony and that harmony is only produced when proportions of things are seen or heard simultaneously?'

As Leonardo constantly searched for a unifying principle, he considered how the laws of nature were rooted in harmony.

'The part always has a tendency to unite with its whole in order to escape from its imperfection.'

The harmony expressed in his paintings was a testimony to the depth of his understanding and appreciation of 'the infinite works of nature' and his total skill in depicting them.

'The work is the first thing born of union; if the thing that is loved is base then the lover becomes base. When the thing taken into union is in harmony with that which receives it, there follow delight, pleasure and satisfaction.'

Love

LEONARDO'S LIFE was one long pursuit of knowledge and his reflections indicate that his quest for knowledge arose out of a profound love of all things.

> 'For in truth great love springs from the full knowledge of the thing that one loves; and if you do not know it you can love it but little or not at all.'

Above: Detail from *St Anne with the Virgin and Child, and John*

As Vasari wrote: 'Leonardo's disposition was so loveable that he commanded everyone's affection. He owned, one might say, nothing and he worked very little, yet he always kept servants as well as horses. These gave him great pleasure as did all the animal creation which he treated with wonderful love and patience.' Despite the restlessness and frustrations in his life, Leonardo always emphasized mankind's essential need for love.

'The lover is moved by the thing loved, as the sense is by that which perceives, and it unites with it and they become one and the same thing... when the lover is united with the beloved it finds rest there; when the burden is laid down there it finds rest.'

'If there is no love, what then?'

Wisdom

IN 1516, FRANCIS I of France persuaded Leonardo to come to his court at Amboise on the Loire. Francis loved Leonardo for the great artist that he was, but above all for his wisdom, writing: ‘No other man had been born who knew as much about sculpture, painting and architecture, but still more he is a very great philosopher.

Although on these two pages are to be found particular words of wisdom, born out of love and rooted in observation, there is little of Leonardo's writing that does not possess the mark of wisdom.

'Ask advice of him who governs himself well.'

'If you governed your body by the rules of virtue you would have no desire in the world.'

'To speak ill of a base man is much the same as to speak ill of a good man.'

'Reprove your friend in secret and praise him openly.'

'Thou, O God, dost sell unto us all good things at the price of labour.'

The soul

THE INNER or spiritual life shone through Leonardo's work. It was unmistakable in his great religious paintings.

'The soul can never be corrupted with the corruption of the body, but acts in the body like the wind which causes the sound in the organ, where if the pipe is spoiled, the wind would cease to produce a good result.'

It is also plain in even the simplest of his drawings. These are more than mere physical representations, they are the product of of 'seeing in depth'; making a connection with life's inner nature.

'A good painter has two chief objects, man and the motion of his soul, the former is easy, the latter hard.'

Whoever would see how the soul dwells within its body let him observe how this body uses its daily home, for if this is without order and confused the body will be kept in disorder and confusion by its soul.'

Ignorance

LEONARDO HAD great ideals. He also recognized mankind's capacity to fall far below its true standard.

'The greatest deception men suffer is from their own opinions.'

'Men out of fear shall pursue the things they most fear: that is they will be miserable lest they should fall into misery.'

Of the cruelty of men, he said:

'Creatures shall be found on earth who will always be fighting one with another. There will be no bounds to their malice... the gratification of their desire shall be to deal out death, affliction, labour, terror and banishment to every living thing.'

Opposite: *Prisoner costume for a masque*

Prophecies

As someone who had studied the physical laws that govern creation, Leonardo was also aware of those natural laws that govern mankind's actions. It was his belief that every deed had its consequence.

Of swords and spears, Leonardo wrote:

'That which of itself is gentle and void of all offence will become terrible and fierce by reason of evil companionship, and will take the lives of many people with the utmost cruelty; and it would slay many more if it were not that these are protected by bodies which are themselves without life, and have come forth out of pits – that is by breastplates of iron.'

Of money and gold:

'Out of the cavernous pit a thing shall come which will make all the nations of the world toil and sweat with the greatest torments, anxiety and labour, that they gain its aid.'

And the fear of poverty:

'A malignant and terrifying thing will spread so much fear among men that in their panic desire to flee from it, they will hasten to increase its boundless power.'

Tales

As a singer, musician, designer and deviser of court entertainments, the ability to tell amusing stories was a required ability for Leonardo. Stories have always been a means of teaching, and Leonardo was undoubtedly a conveyor of wisdom. It was natural enough, therefore, for him to compose narratives that both entertained and informed. And there are a number of such stories to be found in his notebooks. Despite his desire to live a solitary life, his company was always sought after. His talent as storyteller must have formed part of his appeal:

THE ANT AND THE GRAIN OF MILLET

The ant found a grain of millet. The seed, feeling itself caught, cried out, 'If you do me the kindness to allow me to accomplish my function of reproduction, I will give you a hundred such as I am.' And so it was.

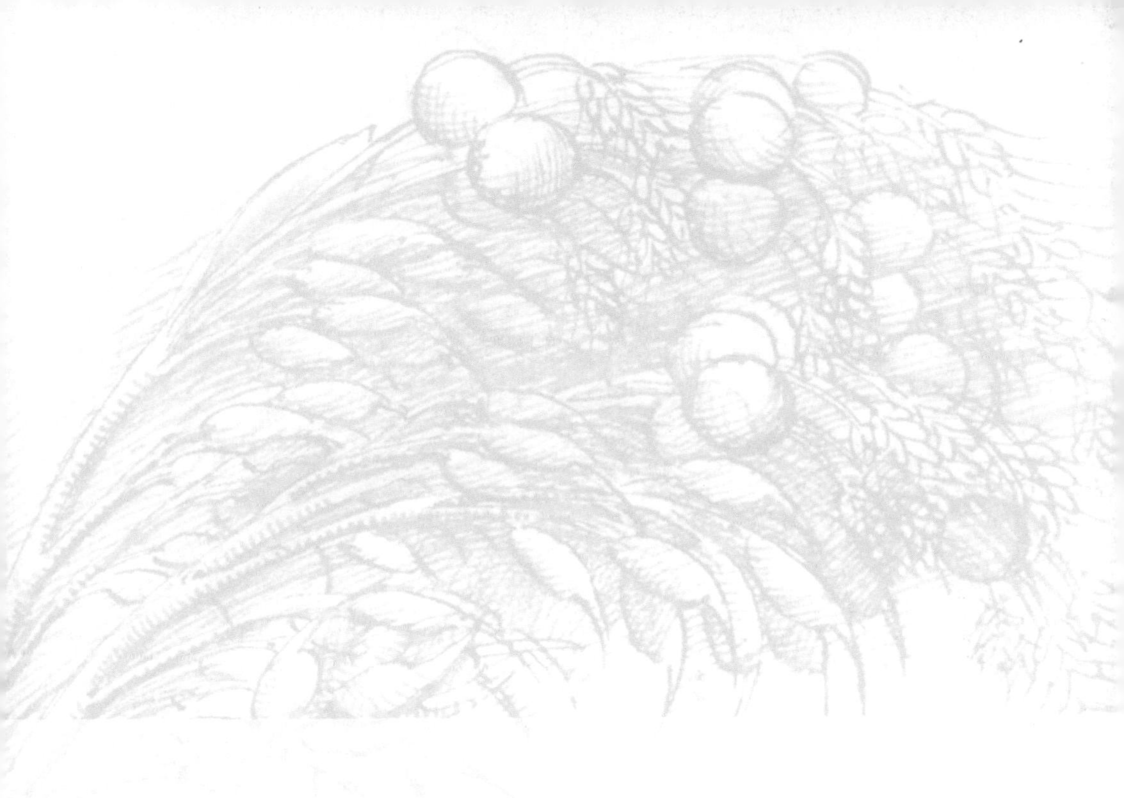

And again:

The Franciscan friars are wont to keep certain periods of fasting when they do not eat meat in their monasteries, but on journeys as they are living on charity they have licence to eat whatever is set before them. Now a couple of these friars travelling under these conditions stopped at an inn in the company of a certain merchant, and sat down with him at the same table, where on account of the poverty of the inn nothing was served except a small roasted cockerel. At this the merchant, as he saw that this would be little for himself, turned to the friars and said, 'If I remember rightly you do not eat any meat in your monasteries at this season'. At these words the friars were constrained by their rule to admit without further cavil that this was the case; so the merchants had his desire and ate the chicken, and the friars fared as best they could.

Now after they had dined thus, the table-companions departed all three together, and after travelling some distance they came to a river of considerable width and depth, and as they were all three on foot – the friars by reason of their poverty, and the other from avarice – it was necessary, according to the custom of company, that one of the friars, being barefoot, should carry the merchant on his shoulders; and so the friar, having given him his clogs to hold, took up the man. But as it so happened, the friar, when he found himself in the middle of the river, remembered another of his rules, and stopping short like St Christopher raised his head towards him and said, 'Just tell me have you any money about you?' 'You know quite well that I have,' answered the other, 'How do you suppose a merchant like me could go about otherwise?' 'Alas,' said the friar, 'our rule forbids us to carry any money on our persons,' and forthwith he dropped him into the water. As the merchant perceived that this was done as a jest and in revenge for the injury he had done them, he with smiling face, and blushing somewhat from shame, endured the revenge peaceably.

And finally:

A man gave up associating with his friend because the latter had a habit of talking maliciously against his other friends. This neglected friend one day reproached him and with many complaints besought him to tell him the reason why he had forgotten such a great friendship as theirs; to which he replied, 'I do not wish to be seen in your company any more because I like you, and if you talk to others maliciously of me, your friend, you may cause them to form a bad impression of you, as I did when you talked maliciously of them to me. If we have no more to do with each other it will seem as though we had become enemies, and the fact that you talk maliciously of me, as its your habit, will not be blamed so much as if we were constantly in each other's company.'

Symbolism

MANY HAVE attempted to interpret the symbolism that is to be found in Leonardo's work. In his paintings of the Christian themes, he included a wide range of symbols which the devout would immediately have responded to. However, many scholars consider that there is much more hidden in these paintings. Engaged as he was in all kinds of artistic ventures, Leonardo was probably called upon to use symbolism extensively in the form of pictorial decoration or for the designs for court masques.

'Patience serves us against insults precisely as clothes do against cold. For if you put on more garments as the cold increases, the cold cannot hurt you; in the same way increase your patience under great injuries and they cannot vex your mind.'

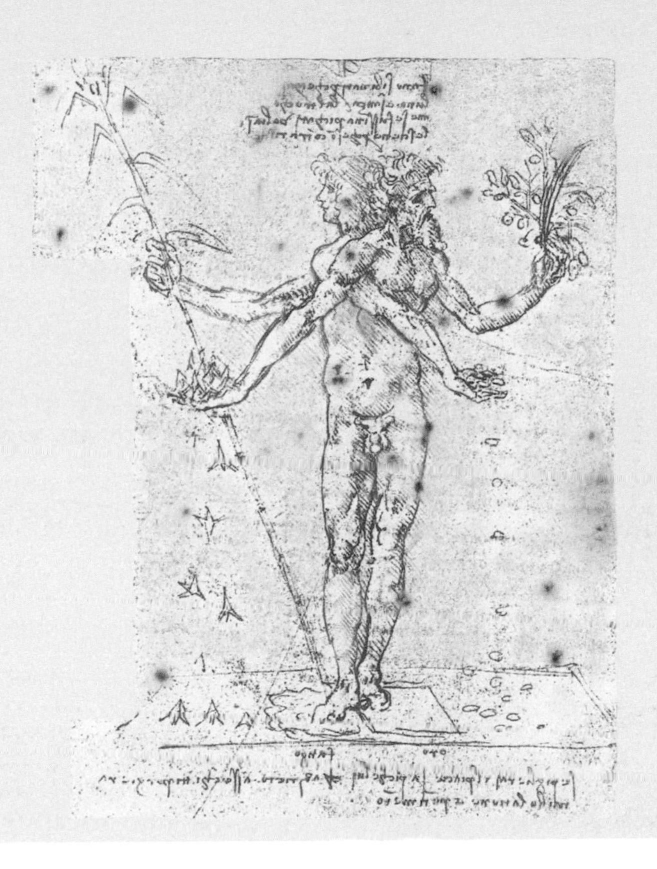

'Pleasure and Pain are represented as twins, since there never is one without the other; and as if they were joined together back to back since they are contrary to each other.'

'Just as iron rusts from disuse, and stagnant water putrefies, or when cold turns to ice, so our intellect wastes unless it is kept in use.'

Reflections on life

LEONARDO portrayed people at all stages of life, and he reflected on the passage of time. He never stopped considering the purpose of life and recording it in all its variety. Like many before him, Francis I of France considered Leonardo a true 'philosopher'.

'Life passes. What is fair passes.'

'We are deceived by promises and deluded by time, and death derides our cares; life's anxieties are naught.'

'He who possesses most is most afraid to lose.'

'The age as it flies glides secretly and deceives one and another; nothing is more fleeting than the years, but he who sows virtue reaps honour.'

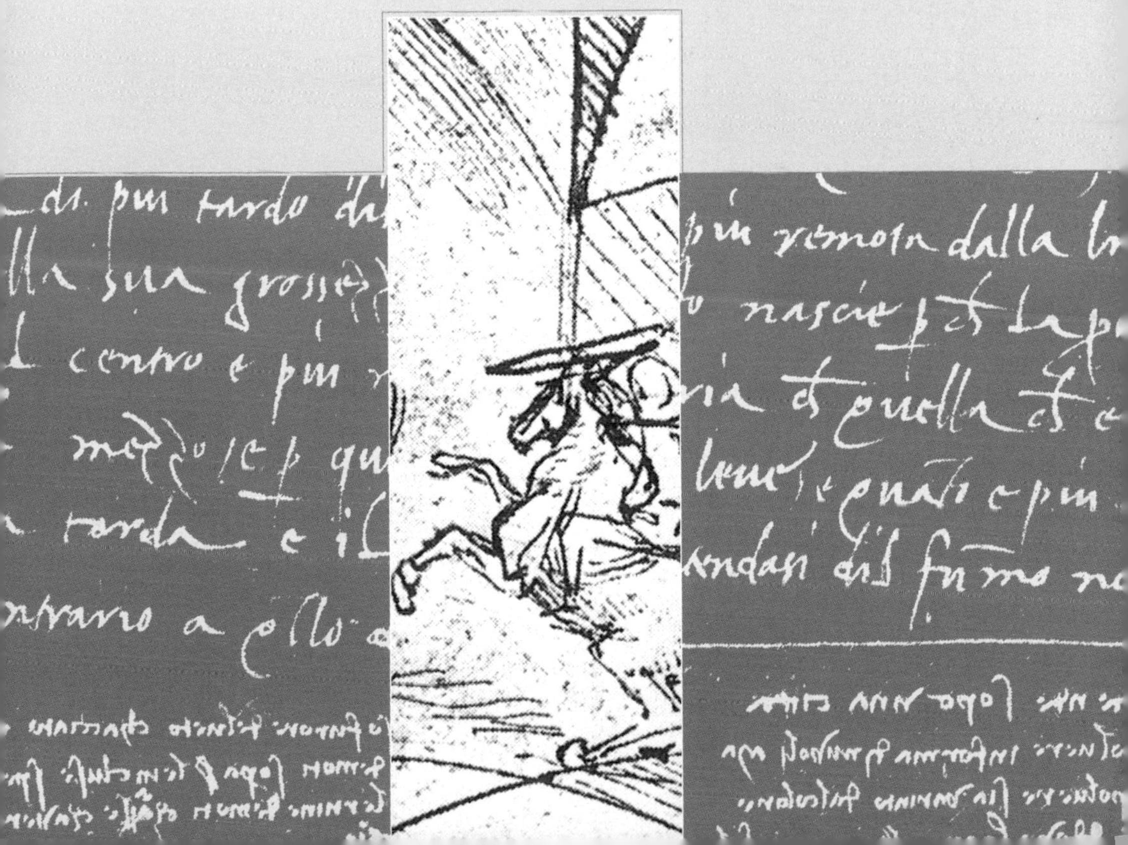

Life of the body

LEONARDO REFLECTED not only on the nature of a good life but also on the nature of a natural and healthy one; a life that was measured and in tune with nature. As ever, harmony was at the root of his thinking. For Leonardo life was full and varied, but also balanced and harmonious.

> 'Medicine is the restoration of discordant elements; sickness is the discord of the elements infused into the body.'

'Nature being inconstant and taking pleasure in creating and making continually new lives and forms, because she knows that they augment her terrestrial substance.'

'The body of anything whatsoever that takes nourishment continually dies and is continually renewed.'

'Nourishment... if it is exhausted it no longer has life.'

'To keep in health this rule be wise:
Eat only when you want and sup light.
Chew well, and let what you take be well cooked and simple.'

'Rest your head and keep your mind cheerful.'

'Shun wantonness, and pay attention to diet.'

The light of truth

LEONARDO WAS in the pursuit of truth and could never be satisfied with falsehood of any kind. Truth was the guiding light of his life: it was mankind's true purpose and inherent in this purpose was mankind's true nobility.

'Truth is so excellent that if it praises but small things they become noble.'

Leonardo was a master of the depiction of the play of light and there is an inner radiance to be detected in his work: for him, truth and light were closely linked.

'To discern, to judge, to reflect are actions of the human mind.'

'Fire destroys falsehood that is sophistry, and restores truth, driving out darkness. Fire is to be put for the destroyer of every sophistry, as the discoverer and demonstrator of truth; because it is light, the banner of darkness, concealer of all essential things.'

'Vain splendour takes from us the splendour of true being.'

'Light is the chaser away of darkness.'

'Nothing is hidden under the sun.'

The final words

IN THE REGISTER where his death was recorded Leonardo was described as: 'painter, engineer, architect and state mechanist', but those were, of course, just a few of his abilities. He constantly pushed the boundary of knowledge, exploring things that didn't find final fruition until hundreds of years later: Leonardo's notebooks were not published until four hundred years after his death, and in them people were amazed to find that what they thought had been the latest discovery had been prefigured by his fertile mind.

The people of his own day, even if they knew little of his private explorations, realized from the very beginning that there was a supremely gifted artist in their midst. His genius was confirmed by the writings and drawings contained in his notebooks. Here is found the kind of creative thinking the human mind is capable of when it works at full capacity. Not least amongst these are insights into Leonardo himself; what motivated him, what aroused his curiosity, what understanding he came to through his acute observations. His understanding of nature and the meaning of life reveal to us why he is still revered as an artist of unsurpassed power, and a thinker who did so much to create many of the hallmarks of the modern age. His executor announced his death with these words:

'As long as these my limbs endure I shall possess a perpetual sorrow, and with good reason… It is a hurt to anyone to lose such a man, for nature cannot again produce his like.'

There were, and are, few who would disagree.

Picture Credits

The Royal Collection, Copyright 2005, Her Majesty the Queen Elizabeth II:
P14 Baby in the womb
P22 Grotesques
P23 Head of Leda
P27 St Bartholomew
P34 Judas
P39 Draped Sleeve
P56 River Scene
P57 Bird's Eye View of Ferry Crossing a River
P60 Star of Bethlehem
P68 Head and Shoulders of Old Man
P69 Muscles of the Back
P61 Copse of Birches (detail)
P76 Two Plants
P77 Arezzo and the Chiana Valley (detail)
P80 Allegory of a Wolf and an Eagle
P82 Head of a Woman
P83 Cats, Dragons etc (detail)
P85 Head of an Elderly Man in a Hat (detail)
P67 Self Portrait of the Artist in Old Age (detail)
P87 Study of Water Passing Obstacles (detail)
P89 Profiles from Adoration of the Magi
P91 St George and the Dragon (detail)
P95 Horse's Head
P96 Hurricane
P99 Earthquake, Fiery Rain and Explosion (detail)
P114-115 Valley scene
P115 Study of Water Passing Obstacles
P116 Arezzo and the Chiana Valley
P130 Cannon Foundry
P134 Horseman Galloping
P138 Skeleton leg bones
P145 Early Tank (detail)
P146 One-man Battle Ship
P150 Allegory of a Wolf and Eagle (detail)
P163 Example of Sheet Music
P164 Oak Leaves (detail)
P165 Study of Arms and Hands
P167 Head of St Anne
P170 Self portrait in Old Age
P175 Prisoner Costume for Masque
P178-179 Heads of Two Types of Rush
P181 Grotesques
P182 Neptune and Rearing Sea Horses
P186 Sprig of Blackberries
P187 Nude back View of Man

Reproduced with the kind permission of Bridgeman Art Library:
P50-51 Sketch for the Adoration of the Magi
P111 Archimedes Screw
P132 Catapult Winch
P140 Fin Spindle

Reproduced with the kind permission of Archivi Alinari, Florence:
P24 and 191 Self Portrait
P78 Arno Landscape
P160-161 Three Women Dancing

Reproduced with the kind permission of Dover Books:
P7 Head of young woman
P26 Organ of Perception, the 'sensus communis'
P28-29 Shouting Warrior
P30-31 St Anne, the Virgin and Child and St John
P32-33 Sketch for a battle scene
P38 Draped knees
P42 Death of a Dragon
P45 Head of a young woman
P46 Sketch for the Adoration of the Magi
P49 Sketch for the Adoration of the Magi II(detail)
P62 Storm
P65 Vetruvian Man
P67 Proportions of a Face
P71 Sketch for a Monument
P81 Sketch for the Madonna, Child and Cat
P86 Sketch for Figures from the Last Supper
P88 Man in Fanciful Armour
P58 Proportions of a Leg
P101 Alphabet from the Notebooks
P105 Hanged Man
P106 Sketch of a Church
p107 Bust of old Man (detail)
P112 Earth in its Layers
P118 Waterfall Sketch
P119 Water Studies
P126 Storm (detail)
P133 Catapult
P137 Cogged wheel (detail)
P139 Allegory of a Skeleton
P152 Angel Placing a Shield on a Trophy (detail)
P169 Sketch for the Madonna, Child and Cat
P173 Sketch for the Madonna and Fruit
P183 Allegory of Pleasure and Pain
P184 Hanged Man: Bernardo di Bandino
P188 Head of young Woman

Reproduced with the kind permission of Art Archive Picture Library:
P36 St Anne with the Virgin and Child and St John
P40 Seated St Anne, Virgin and Child
P47 *Angel from* The Virgin of the Rocks
P73 Mona Lisa
P75 Adoration of the Magi
P93 St John the Baptist
P123 The Annunciation
P146 Scythed War Machine
P149 Leda and the Swan
P155 Helicopter
P168 St Anne (detail)

Copyright Biblioteca Ambrosiana:
P144-145 Huge Mortars with Exploding Projectiles
P148 Sketch of Wing Design

Any unlisted images, decorations or motifs are also reproduced with the kind permission of Dover Books